Atlas of the History of Modern Science 1500-2020

Thomas H. Brobjer

Atlas of the History of Modern Science 1500-2020

Overviews and Surveys in the Form of Charts, 'Maps' and Diagrams

Springer

Thomas H. Brobjer
Department of the History of Science & Ideas
Uppsala University
Uppsala, Sweden

ISBN 978-3-031-76626-8 ISBN 978-3-031-76627-5 (eBook)
https://doi.org/10.1007/978-3-031-76627-5

This Springer imprint is published by the registered company Springer Nature Switzerland AG
The registered company address is: Gewerbestrasse 11, 6330 Cham, Switzerland

To Marika, Max, and Mattias

Acknowledgment

The work on this atlas has been guided, stimulated, and profited in many ways by discussions with friends and many students. I would especially like to mention my father Hans, Mart, Max, my doctoral supervisor at University of Sussex, John Murrell, Philip, Anna, Gunnar, Magnus, Hanno, and Michael.

Contents

List of Charts

4 History of Physics

5 History of Chemistry

1 Introduction

This work constitutes an attempt to present the history of modern science graphically by means of charts, tables, 'maps', tableaus, flowcharts and overviews. Expressed differently, what is attempted here is a cartography of the history of science. This approach gives overviews and outlines, and emphasizes contexts and relations (including influences and oppositions), and connections between different scientists and between traditions. Furthermore, it makes these contexts and connections almost instantaneously visually apparent – close contexts and outlines but often also broader and more long-term contexts and overviews, which indicate the more general trends in the history of science.

Facts and information is not knowledge and does not necessarily imply understanding. An essential ingredient of knowledge and understanding is contextual, the ability to relate facts and theories to other facts, themes, contexts and interpretations of them. It is as an aid to such contextual understanding that these charts, tables and maps primarily have been constructed. It seems to me that today, in this era of mass-media and internet, more than ever before, it is contexts and broad overviews and outlines that are most needed and helpful – for information about specific scientists, movements or events can easily be acquired, while the broader picture is often difficult to acquire – it requires both extensive reading and scholarly expertise. It is too easy to get drowned by all the details and thus to fail to see the broader picture. Without a wider view one is apt to get lost, lose interest or be misguided. For many themes and questions, a single chart or 'map' giving an outline, just like a map or a picture, can be worth many words, and it can, at a glance, give reasonably accurate overviews of scientific landscapes and the history of how we have arrived at them.

What is most emphasized in this atlas is not the thoughts of each individual scientist, but how one (and each) relates to and belongs to a wider context of other thinkers. Especially for the broader and more surveying charts and maps a scientist is often placed together with a group of other scientists who thought in a similar manner, and equally important, placed against those who were regarded as the opposition and whom the first set of thinkers attempted to argue against, and in relation to whom they often defined themselves. To understand a scientist, or a group or tradition of thinkers, it is often essential to understand who and what they opposed and argued against. Double-pointed arrows (↔) are frequently used in the charts and maps to indicate such opposition. In many charts and 'maps' scientists and thinkers are also divided into opposing groups or traditions to further emphasize such "negative" influence for determining or directing the orientation of intellectual and scientific positions and traditions.

The advantage of the approach used in this atlas, with charts, 'maps' and tables, is in many ways just the opposite of that of traditional texts, and can thus be complementary to the reading of conventional books and treatises about different aspects of the history of science. Charts and maps are well suited for structuring, organizing and presenting information, especially when one wishes to emphasize outlines, relations, contexts and/or influences and oppositions between the things charted out. Most of the charts, tables and 'maps' in this book are multidimensional. Many of them are four dimensional in respect to the information they present; chronological, thematic, rating the importance of the scientist in regard to the question or field discussed (expressed by size and differential boldness of print) and showing affinities and tensions between scientists, schools or traditions (often expressed by arrows and double-pointed arrows). Where necessary, rather than adding still more information to most charts, I have instead added further, often connected, charts resulting in short series of charts or tables on different related topics. Other charts are more thematic.

The charts and 'maps' contained in this atlas can be used in a manner not unlike how we use maps in navigating and

T. H. Brobjer, *Atlas of the History of Modern Science 1500-2020*, https://doi.org/10.1007/978-3-031-76627-5_1

finding our way in or through a landscape, and as a guide to finding new interesting spots and sites for setting up camp. It makes a great difference when visiting a new city or territory whether one has access to a map or not. It is my belief that these charts and 'maps' likewise will make it easier to get a better grasp of many of the fields of science and of our present worldviews. They also present a skeleton of facts, which aids in structuring knowledge about science and the history of science.

In some limited sense these charts and 'maps' can be regarded as humanistic versions of Feynman diagrams – the graphical method of representing the interactions of elementary particles. Like such diagrams, these charts serve the purpose of visualization, with the vertical axis almost always representing time. However, these charts are, as to be expected, much less exact, and, of course, serve no mathematical purpose. The horizontal axis represents space in Feynman diagrams, while in these charts it usually represents different views and interpretations of reality – that possibly can be regarded as interpretative space. The most relevant similarity, however, is the task of aiding in giving graphical conceptions of historical change, contexts and development.

Just like a geographical or historical atlas, this work is not meant to be read from beginning to end, but rather it is intended to be used for its graphical presentations of overviews of questions, ideas, fields or periods of the history of science. The book aims to give bird's-eye-views and bring forth contexts and the main trends in the history of science. Science has greatly shaped how we see and view the world. One of the best approaches to understanding science and the modern *Weltanschauung* is to approach it historically.

Sometimes there are more than one 'map' or chart describing essentially the same phenomenon or development in the history of science, for the understanding of many phenomena gains by being viewed from several somewhat different perspectives (just as one can present different sorts of standard maps over the same area, such as for example, political, geological, population and economic maps). In particular, many charts with overviews over long periods are often first presented in a conceptual scheme, where more recent periods are enlarged to allow for more text, and because that which is nearer feels larger and more relevant. Thereafter the chart is often repeated in correct chronological proportions to show the true historical development.

The dates listed after many scientists in the charts are, of course, only approximate and serve primarily to aid in making the chronological context apparent. Most frequently they are the closest even year to when the person was about 40 years old, but sometimes I instead use the year the person published an important or relevant book or made a discovery.

Ideas, values and interpretations of the world, just like intellectual movements and traditions, have no sharp beginnings, ends or other "borders", and often they begin and end at different times in different countries and cultures. Thus all such "borders" in this atlas must be taken as highly approximate. The atlas almost exclusively concentrates on the traditional natural sciences, physics, chemistry and biology, and their subfields. Mathematics and medicine, for example, only have a limited presence. Furthermore, it also concentrates on Western contributions to science. It goes without saying that many of the 'maps', graphs and charts in this book could have been constructed differently, just as conventional books on the history of science are somewhat different from one another.

What is presented in this book can be regarded as the skeleton of the history of science. Much further information needs of course to be added for it to become a body of knowledge, but the charts give outlines into which such further information can be added and they enable the user to get a grasp of contexts and overviews of how ideas and thinkers are related to one another. In many ways it is similar to how we use ordinary geographical maps or an atlas, useful for localizing a place, for determining its distance from other places and for saying something about its general characteristics, but an atlas is usually not the appropriate place to find detailed in-depth information about it. Such detailed information requires further and supplemental reading in conventional works on the history of science. The tableaus contained in this atlas present approximate outlines and overviews. More detailed knowledge will often show how much more complex reality and history has been. They are thus often suitable starting points, and first approximations, but not the end points and conclusions in the understanding of science and the history of science. For example, although many individual names are mentioned, there are, of course, a very large number of names and institutional contexts that cannot be listed in an overview of this kind.

Although many charts are chronologically proportional, there is a sense in which all the charts, especially the long-term ones, are disproportional. There are so many more people working with science today and in the past century that when we compare with the number of people working with science, or the scientific discoveries made, the boxes for antiquity, the Middle Ages and the Renaissance would be miniscule if completely proportional. I almost always use rectangles to represent a tradition or school of thought, etc, but for this same reason, they ought really to be triangles, or at least boxes with very much smaller base in the past times than in the more recent times.

Like the case with most atlases, I have attempted to keep the amount of text to an absolute minimum.

An alternative way to view this present atlas is to see it as an attempt to produce 'pictures', in the form of charts, tables and 'maps', to make it easier for the mind to grasp, to organize and to remember large amounts of complex contextual information.

It has *not* been my intention to write an easily accessible summary of the history of science (even though I am aware that many of the charts, tables and maps can be used as such), but rather to chart many aspects of the history of modern science to make it easier to comprehend, to orient oneself and to navigate among the myriads of views and interpretations of the world which is part of, and have been part of, the many *Weltanschauungen* and worldviews that constitute our intellectual heritage and that influence how we interpret the world today.

There is more than one way of cognition, of thinking and communicating. We do not think and learn using merely verbal ability. Other abilities, such as visual and spatial thinking, including the sense of direction, is also involved in all learning, but especially in the humanistic fields, including history, this has usually been given little room and importance. Many great thinkers, for example, Faraday, Helmholtz, Maxwell, Einstein, Feynman, Hawking and Penrose have emphasized the importance of spatial thinking for them. An atlas of this sort, using graphical representations, utilizes and gives room for and stimulus to also our spatial ability.

Disseminating knowledge and insights with words has many advantages, but also some disadvantages. It is often slow, inefficient and essentially one-dimensional (that is why we speak of the red *thread* of a text, and the reason natural scientists often prefer tables, graphs and mathematics), and it requires and encourages obedience in the reader or listener – if one does not follow the line of thought in the text one has not read the text. Achieving a two-dimensional "flat" communication, not to speak of a three-dimensional complex one, requires a large number of words. On the other hand, the approach of communicating knowledge and insight spatially, involving the sense of direction, using charts and maps, is fast, efficient and inherently two- or three-dimensional. It is probably less suitable for details, but more so for making surveys and giving overviews and outlines.

In most fields, if not all, text is enriched by a suitable addition of pictures. How poor, and how infinitely much longer, would works in art history have to be without pictures. The charts and 'maps' presented in this atlas serve in part a similar purpose to that of pictures. The study of the history of modern science is significantly enriched by the use not only of words but also of visual and spatial presentations.

Map Legend

The chronology in the maps and charts goes downward, i.e. the chronology follows how we normally read, from top to bottom (sometimes from left to right), from older to more recent.	
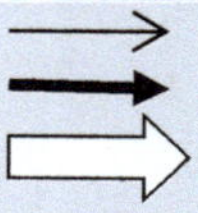	Influence. Sometimes the degree or strength of influence is indicated by the boldness and/or size of the arrow.
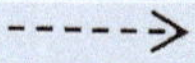	Weak or uncertain influence.
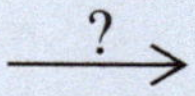	Uncertain influence.
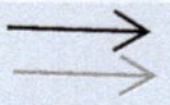	A lighter, grey, arrow indicates a weaker influence. However, a lighter colour is also sometimes used to avoid to clutter up dense parts of maps or charts.
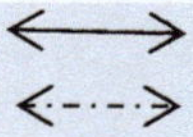	Opposition, or sometimes simply difference. In charts with many arrows, these double arrows are often slashed to more easily distinguish them from (positive) arrows.
	Connection between thinkers or movements, where there is no obvious influence from one to the other, but shared views.
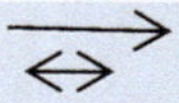	Influence, but also with important opposition or critique. (When the arrow is longest.)
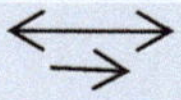	Opposition, but also with important influence or similarity. (When the double-pointed arrow is longest.)
	A group of similar thinkers, traditions or events.
	A dotted or slashed square indicate that it is less important, or sometimes that it represents something alien or foreign in regard to the theme or chronology in the table, e.g. the relevance of philosophical development among a map which charts scientific development.
	In some maps, one or several thinkers or movements are placed in boxes. The length of these boxes generally corresponds to the period the person(s) or movement(s) was active, and the width to his or their influence or importance.
Huygens Kepler **Newton**	The importance and influence of thinkers in regard to the theme or area covered are indicated by size and boldness of print.
Bruno *Boyle*	To separate thinkers who belong to different traditions, schools or who traditionally are classified as different, such as scientists contra philosophers, the names of one such group have sometimes been printed in italics while those of the other in normal print.
[Darwin]	A name in bracket indicates that the person or movement is important or influential, but is chronologically outside the period covered.

Overviews of the History of Science

2

T. H. Brobjer, *Atlas of the History of Modern Science 1500-2020*, https://doi.org/10.1007/978-3-031-76627-5_2

Map of the Major Scientific Schools of Thought: 1. Outline
Chronologically Proportional

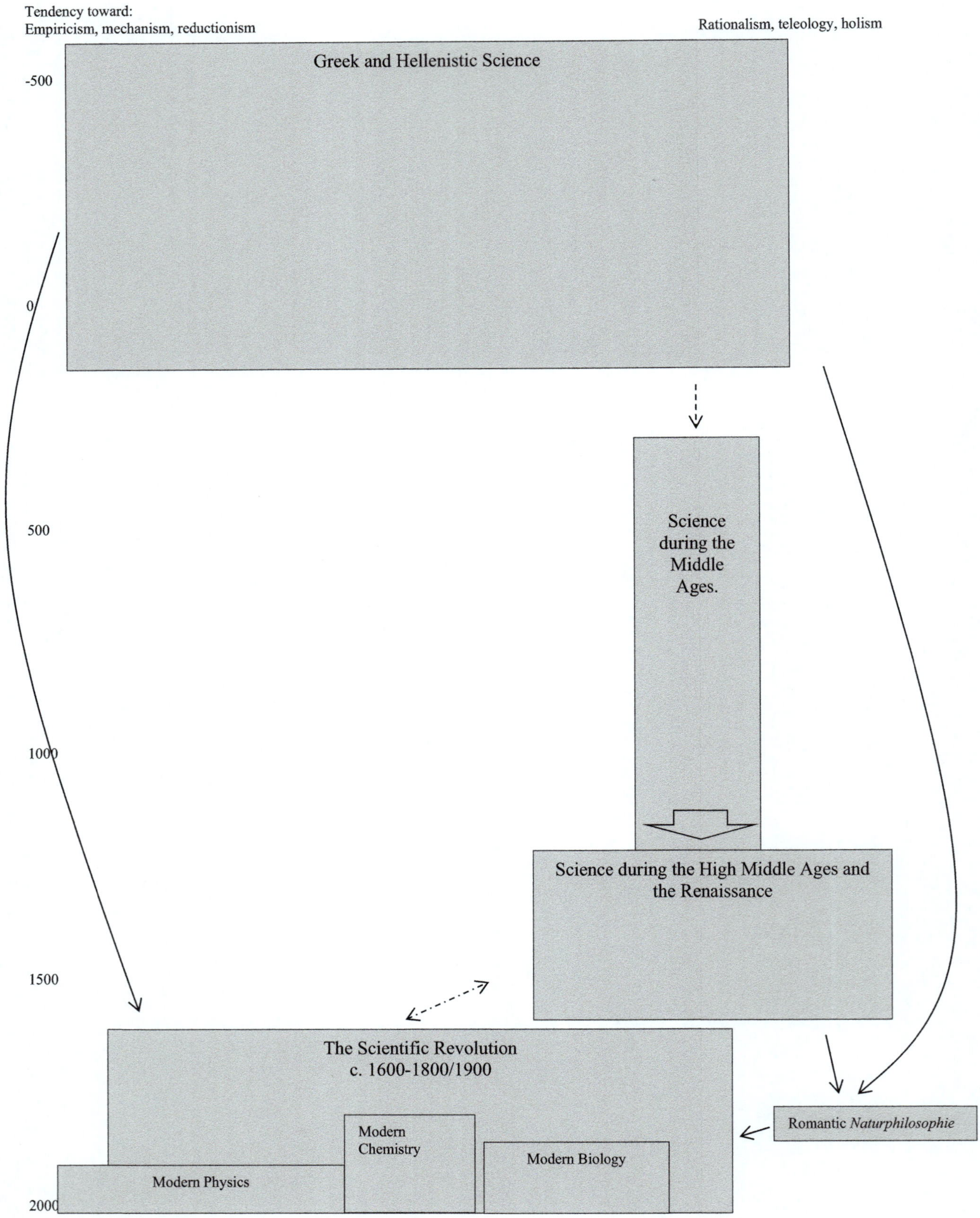

Map of the Major Scientific Schools of Thought: 2. More Details

Chronologically Proportional

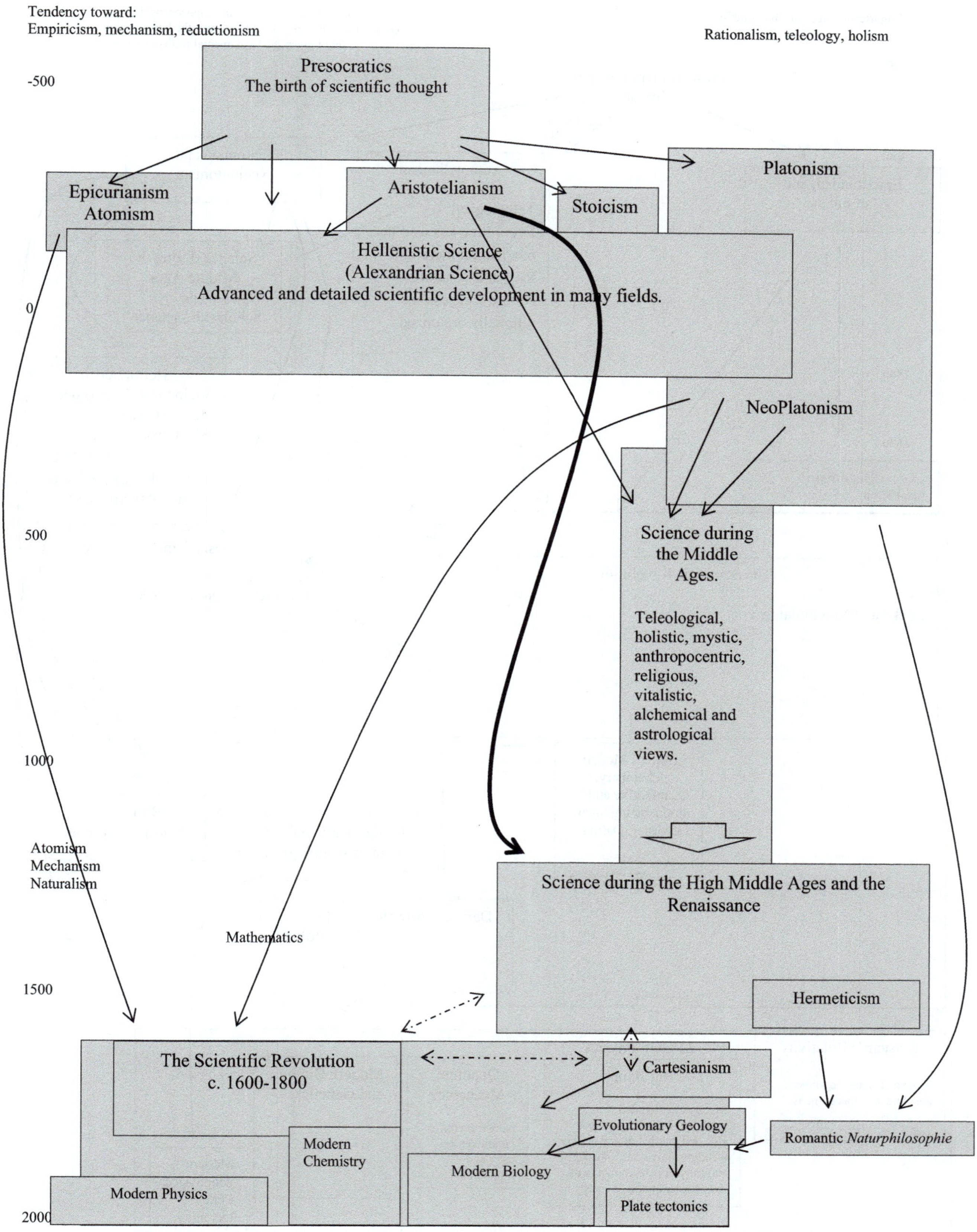

Map of the Major Scientific Schools of Thought: 3

Not Chronologically Proportional (the Modern Era Expanded)

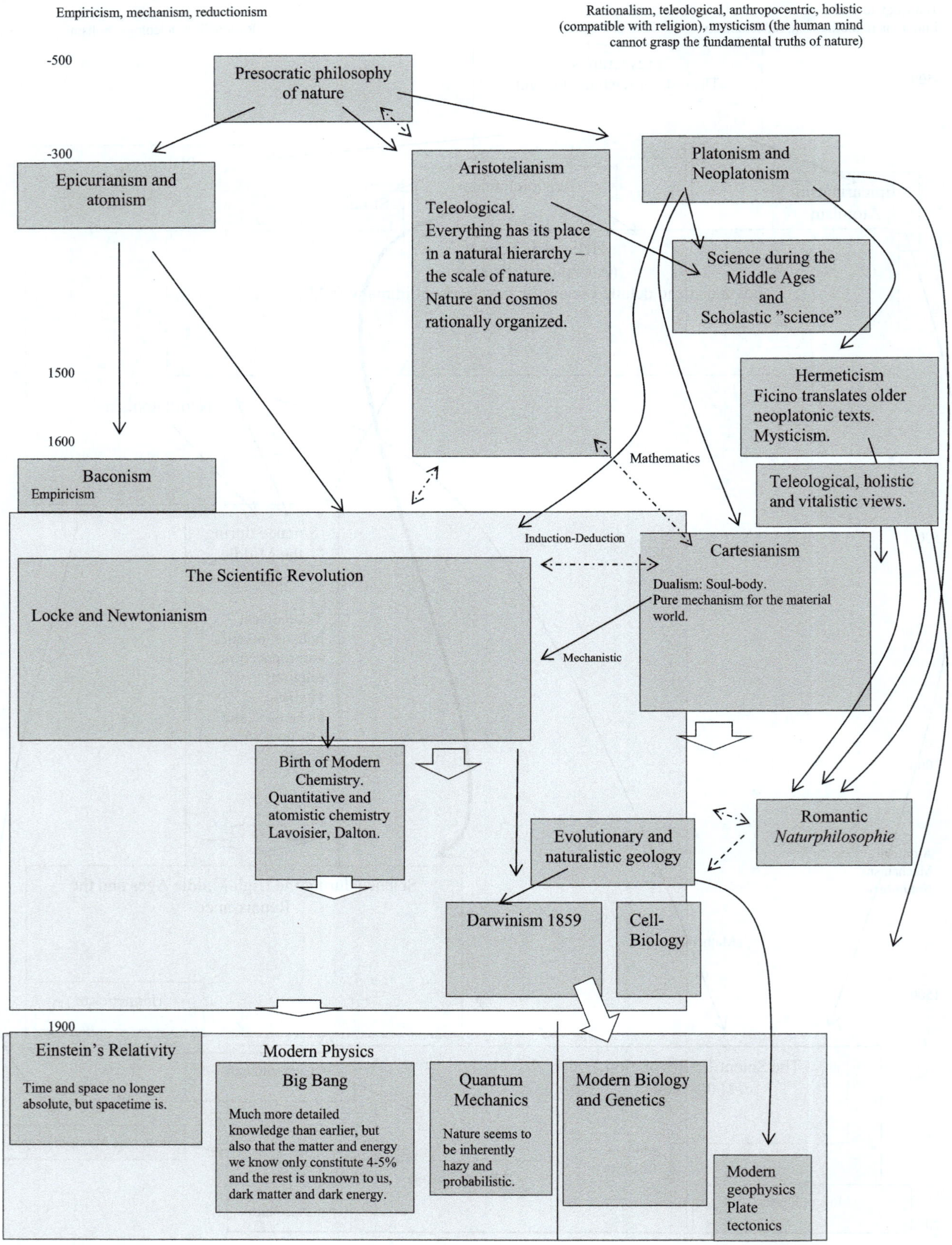

Background Map of the History of Science, 600 B.C. – A.D. 1600

Chronologically Proportional

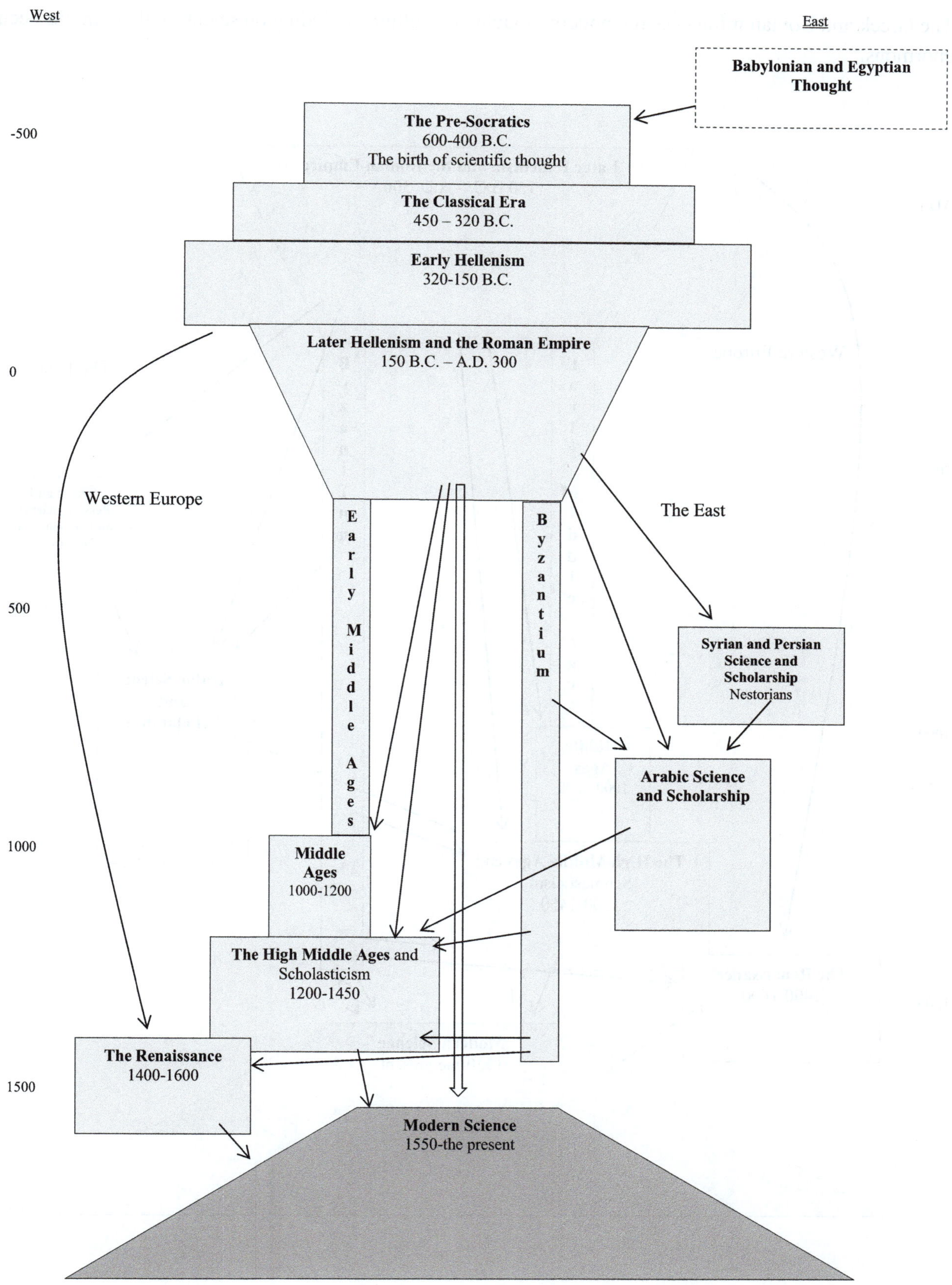

Map of Cultural and Scientific Influences, 150 B.C. – A.D. 1600: 1. Overview

Chronologically Proportional

The Greek and Roman influences on modern thought and culture, including on scientific thought has been enormous.

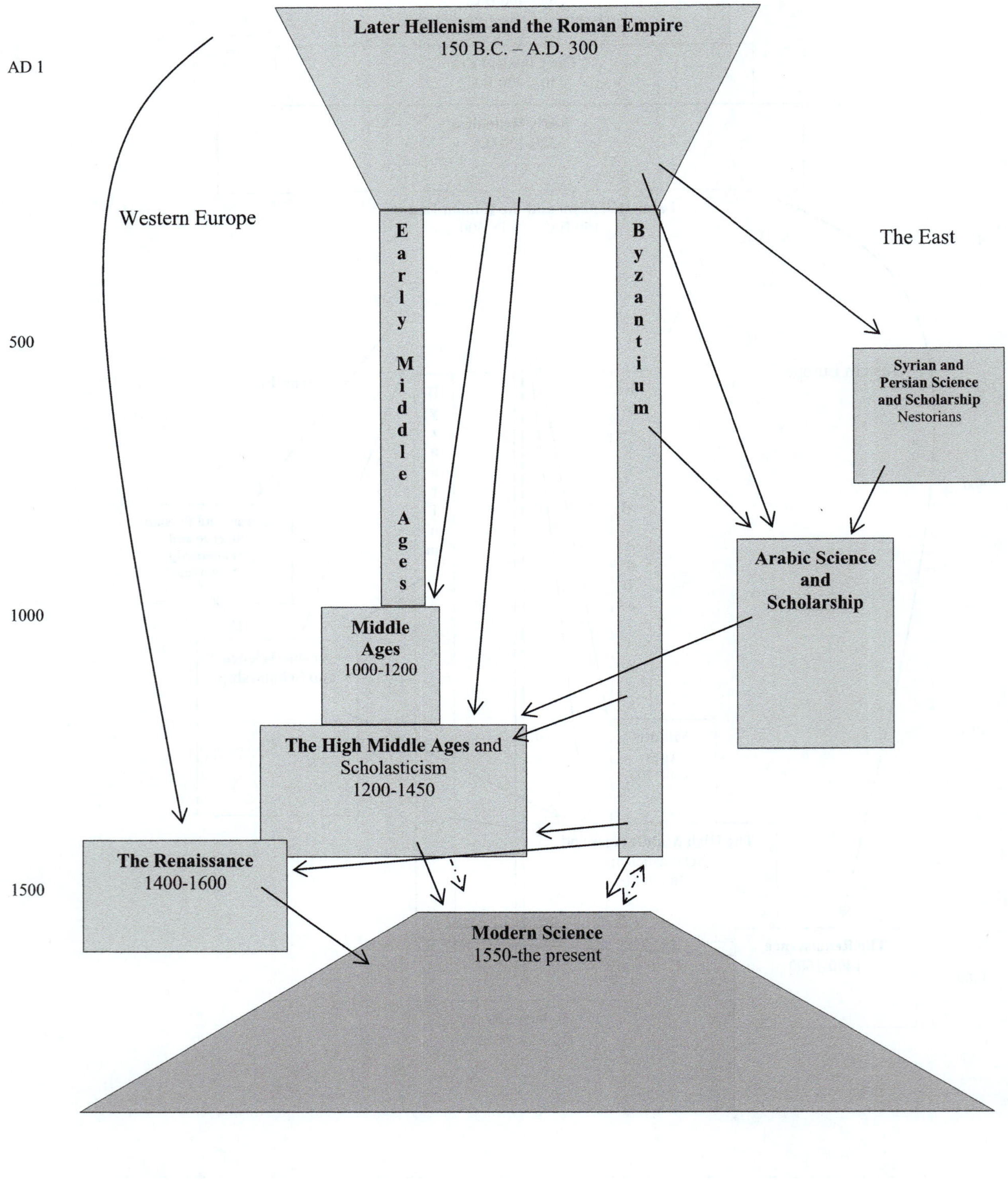

Map of Cultural and Scientific Influences, 150 B.C. – A.D. 1600: 2. With Names

Chronologically Proportional

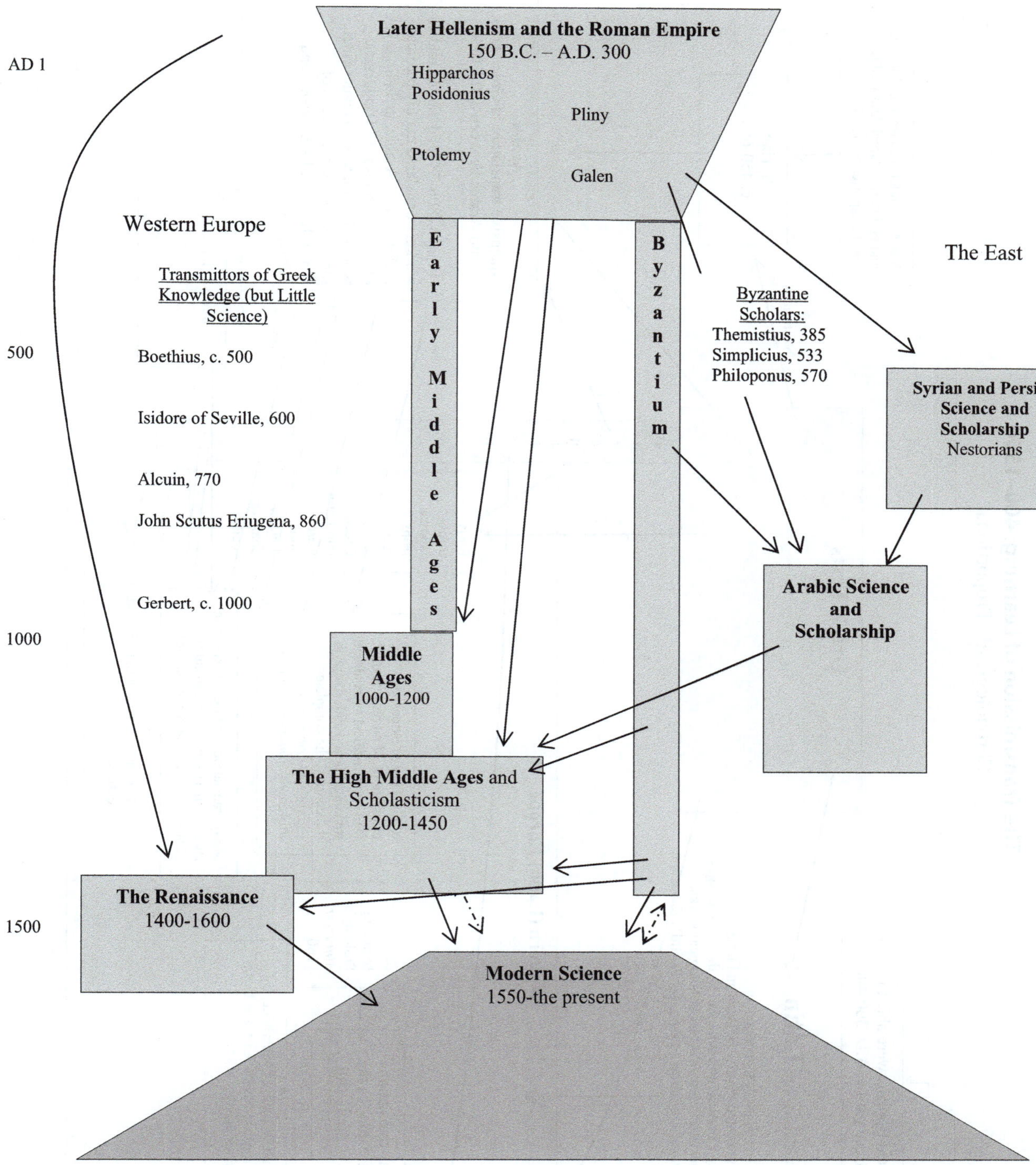

Map of the Translations of Greek Scientific Texts to Other Languages and Cultures (Especially Latin)

The Transmission of Learning: 400–1500
Chronologically Proportional

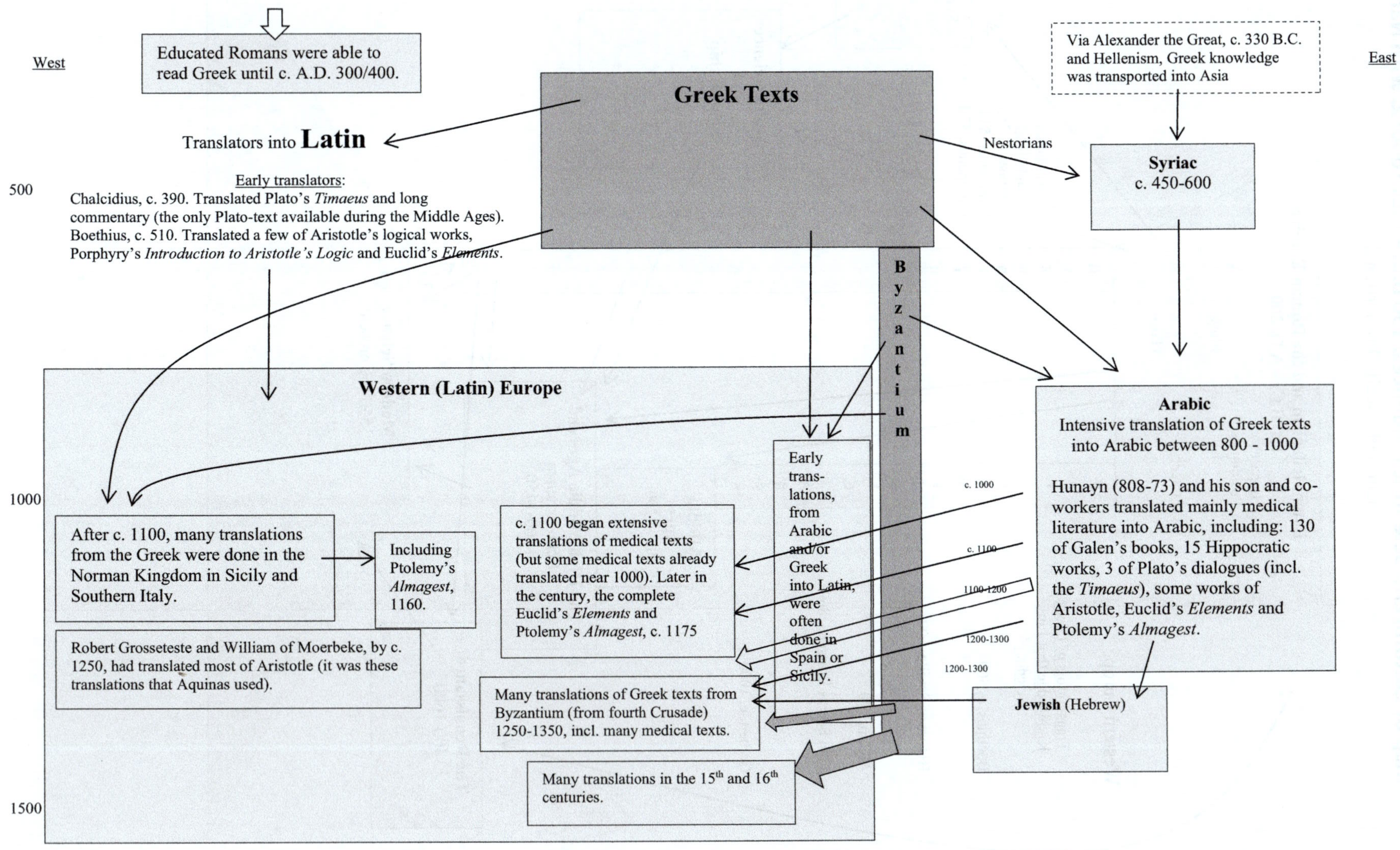

Ancient and Modern Science: General Characteristics

Ancient science was mostly based on observation and theory. They did some experiments (e.g. Strato and Archimedes) and attempted to describe nature with mathematics (e.g. Plato, Callipus and Ptolemy). However, it was first in the seventeenth century that experiment began to be an integral part of the study and understanding of nature, and it was first then that one also began to be able to describe nature accurately enough to formulate mathematical laws (e.g. Kepler, Boyle and Newton).

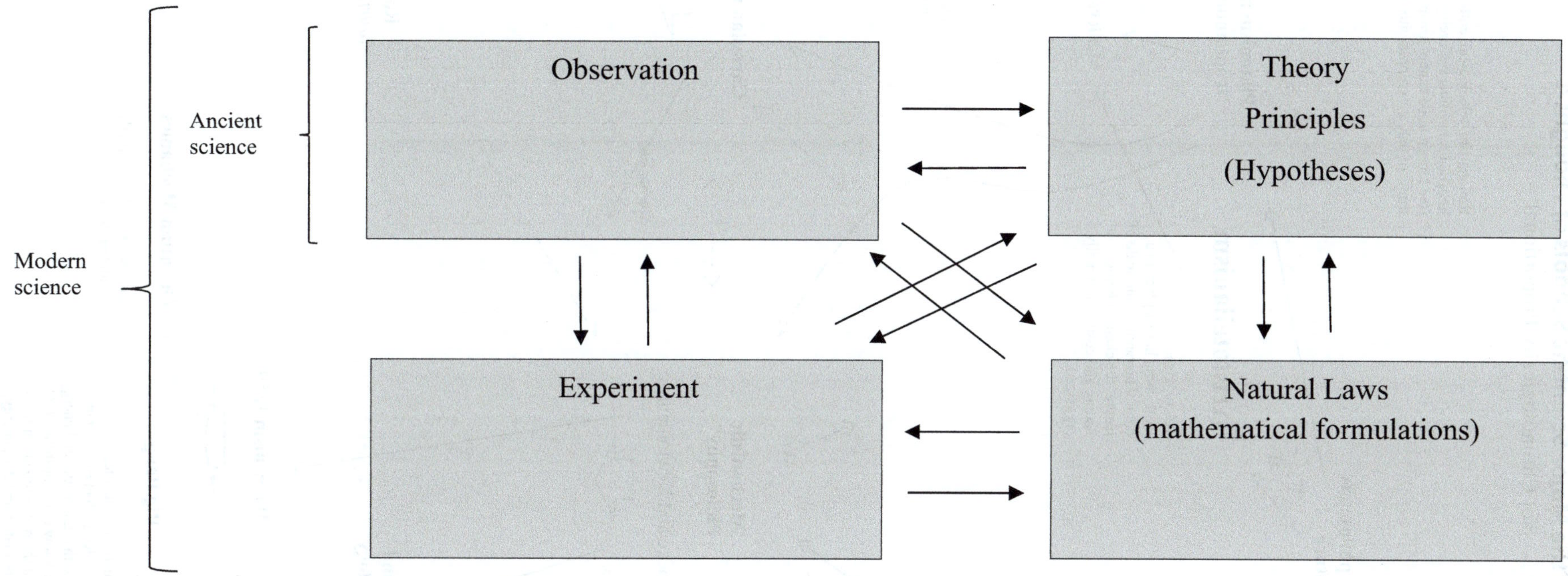

Map of the Major Scientific Schools of Thought 4

Not Chronologically Proportional

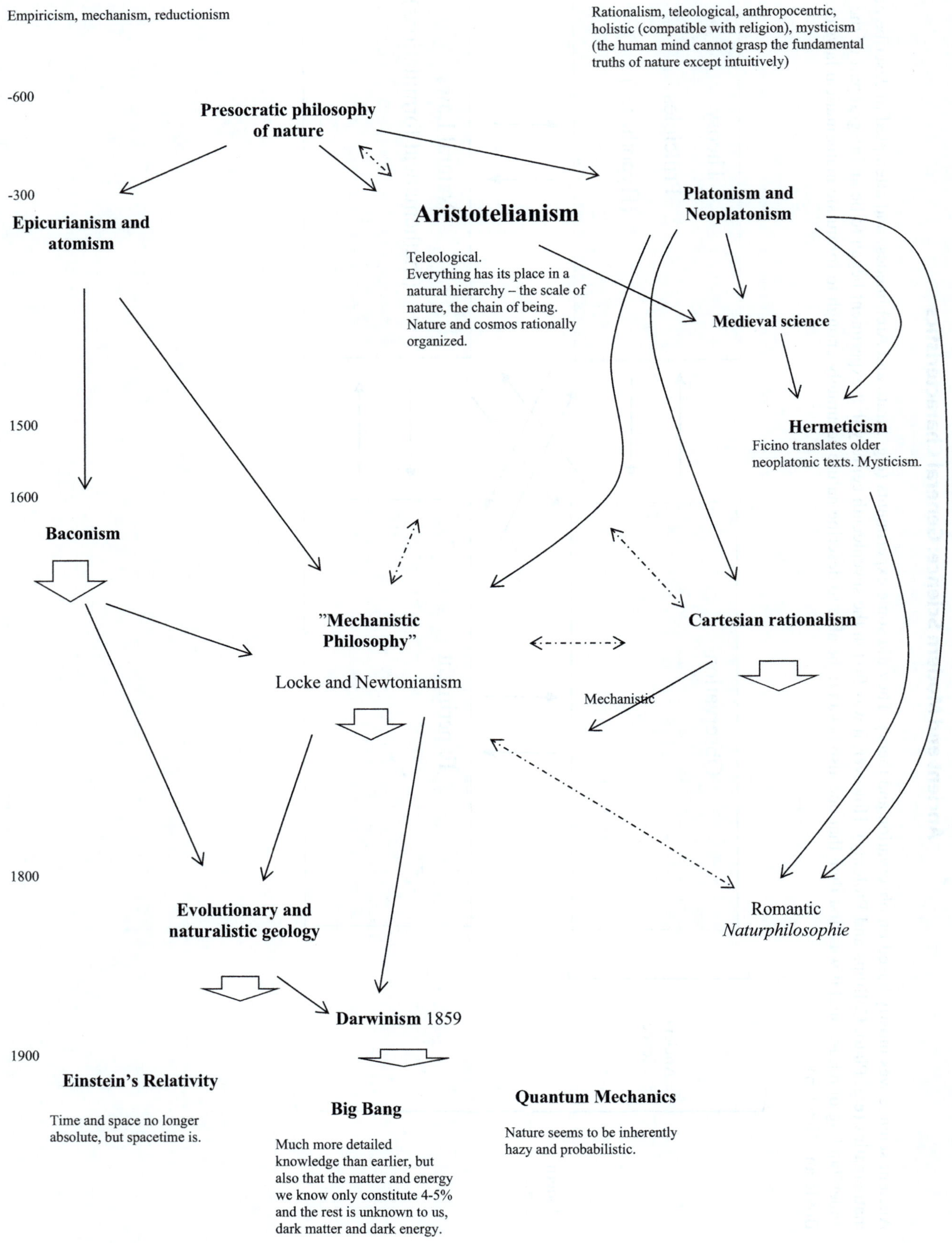

Overview of the Main Epochs of Western Intellectual History

Not Chronologically Proportional

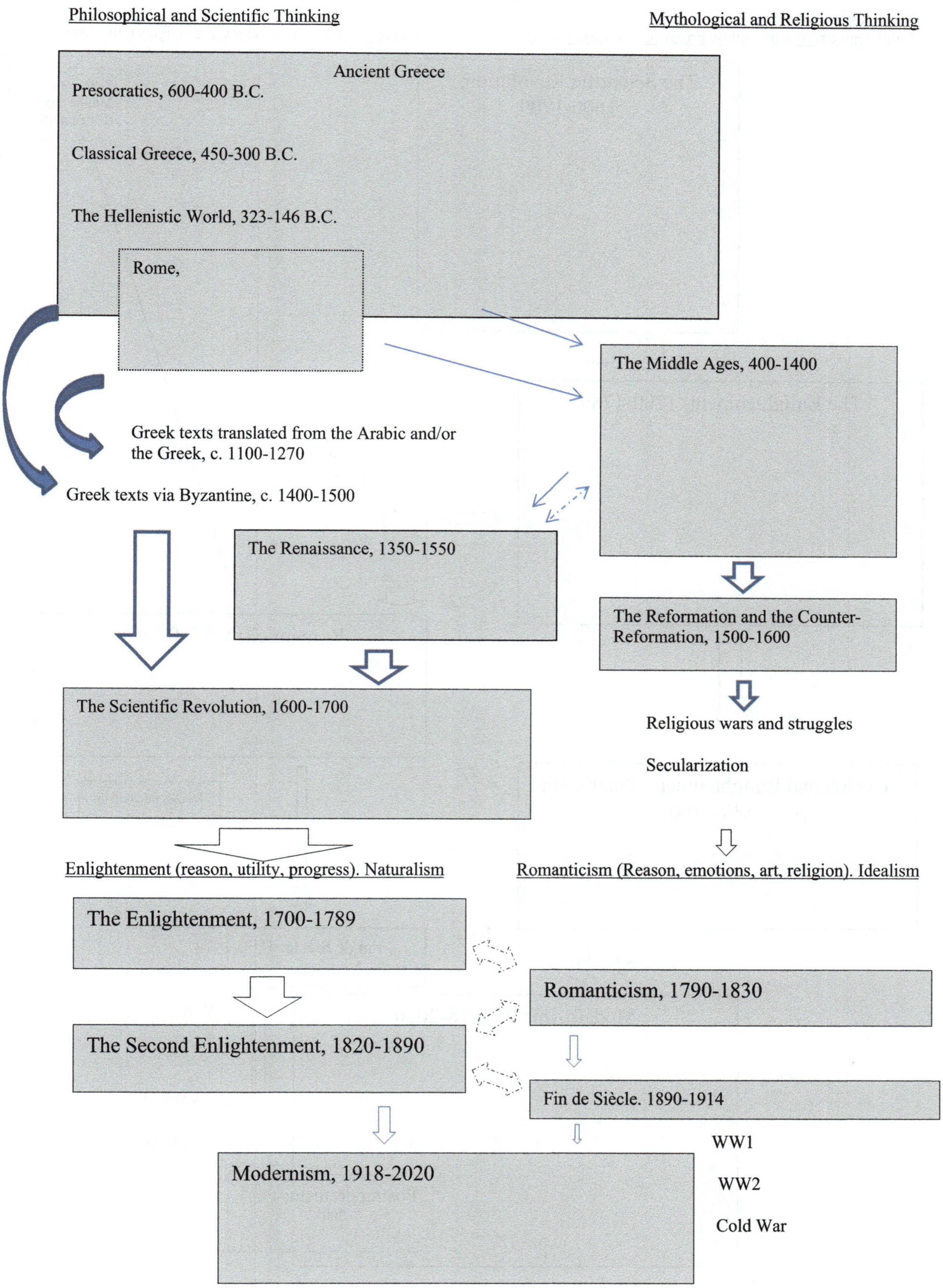

Map of the Modern Epochs in Intellectual History: 1. Post-1600

Chronologically Proportional

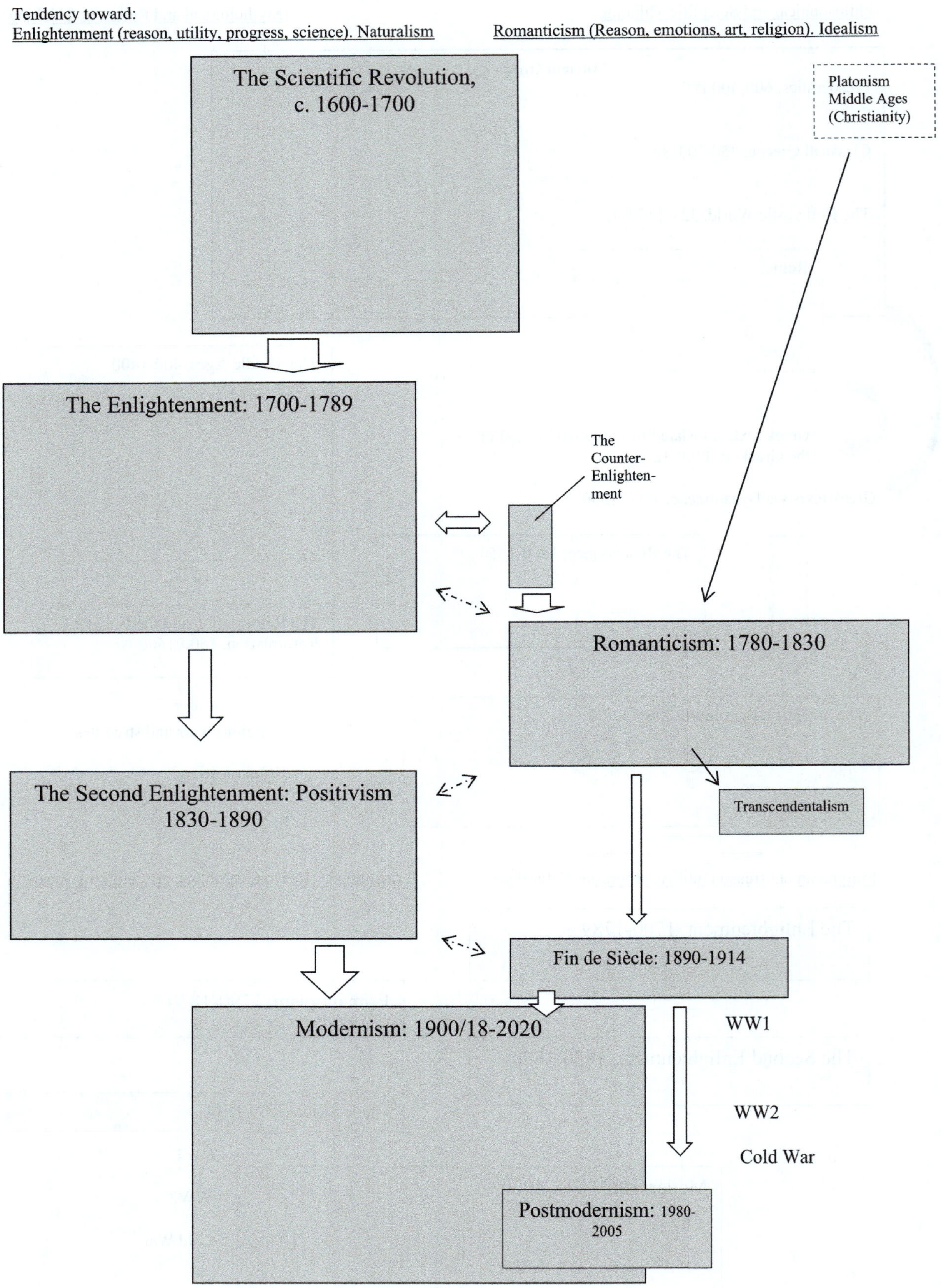

History of Science 1500–2000: Overview

Chronologically Proportional

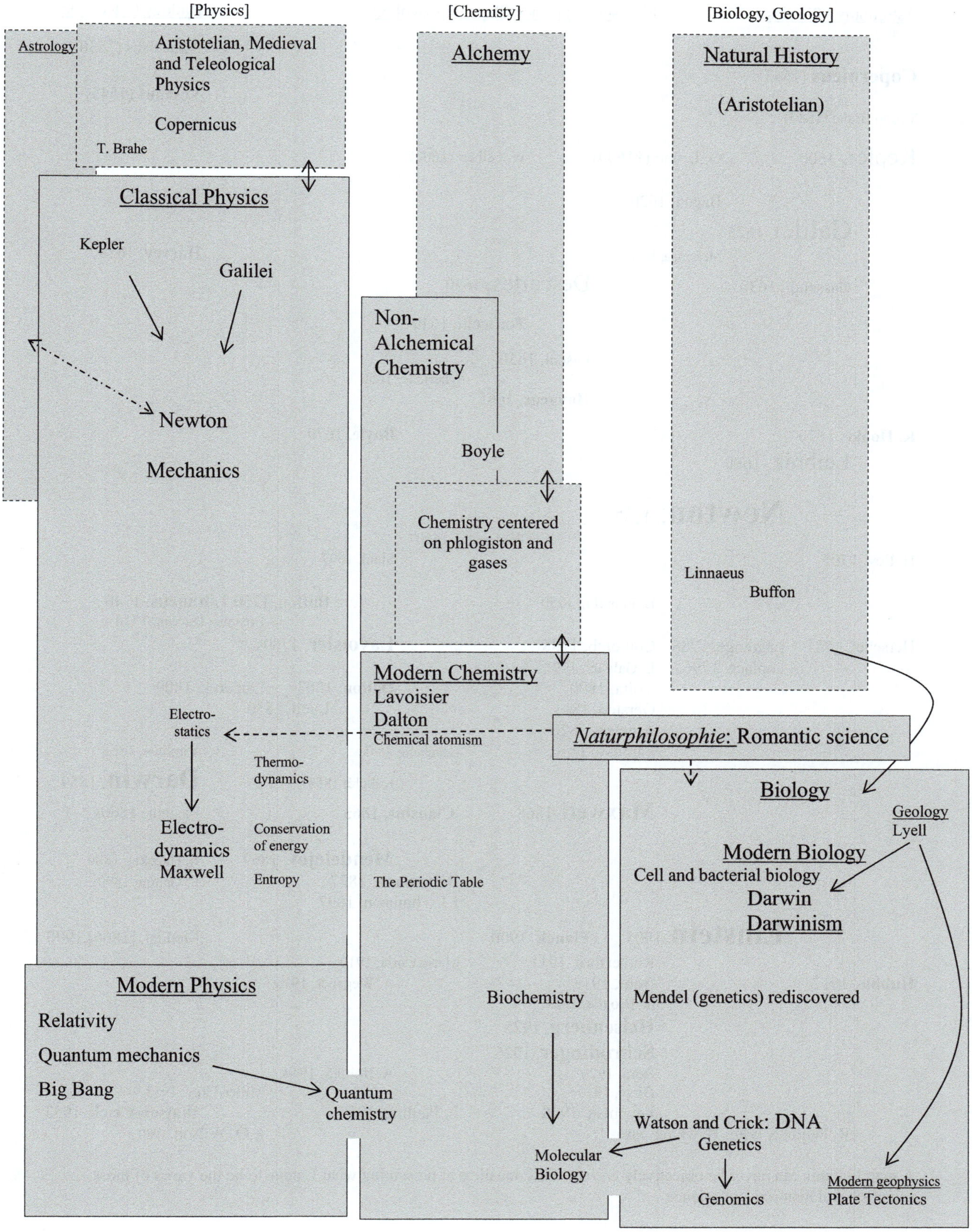

History of Science 1500–2000

Topological Map of Major Scientists

Astronomy / Physics — Philosophy/Physics/Chemistry/Geology — Biology / Medicine

1500

Paracelsus (1530)

Copernicus (1543) **Vesalius** (1543)

Tycho Brahe (1580)

Kepler, 1600 G. Bruno (†1600) W. Gilbert (1600)

Bacon, 1620

Galilei, 1625

Mersenne, 1630 **Harvey**, 1628

Gassendi, 1630 Descartes, 1630

Torricelli, 1640

Pascal, 1650

Guericke, 1650

Huygens, 1665

R. Hooke, 1670 **Boyle**, 1670

Leibniz, 1680

Newton, 1687

Fontenelle, 1690

Halley, 1705 Stahl, 1697

B. Franklin, 1750 Buffon, 1750 **Linnaeus**, 1740

Erasmus Darwin, 1770

Herschel, 1781 Lagrange, 1788 Coulomb, 1788 **Lavoisier**, 1780s

Laplace, 1796 L. Galvani, 1791

Volta, 1800 **Dalton**, 1803 Lamarck, 1809

Oersted, 1811 Lyell, 1830

Ampère, 1820

Faraday, 1830 Helmholtz, 1845 Virchow, 1858

Kekulé, 1858 **Darwin**, 1859

Maxwell, 1868 **Clausius, 1865** Pasteur, 1860s

Mendelejev, 1869 E. Haeckel, 1880

Boltzmann, 1877 F. Galton, 1890

J.J. Thomson, 1897

Einstein, 1905 **Planck**, 1900 Mendel, [1866],1900

Rutherford, 1911 Marie Curie, 1910

Hubble, 1917 Bohr, 1913 A. Wegener, 1915

de Broglie, 1923

Heisenberg, 1925

Schrödinger, 1926

Born, 1926 A. Holmes, 1944

Dirac, 1930 Miller-Urey, 1953

Feynman, 1948 L. Pauling, 1950 **Watson-Crick**, 1953

R. Penrose, 1970 Hawking, 1980 E.O. Wilson, 1980

A map like this can never be objectively correct. This one aims at presenting what I judge to be the views of most scientists and historians of science.

The Relation Between Physics/Natural Science and Culture/the Humanities

Developments in the natural sciences are probably the main cause of changes also for the humanities and general culture.

Classical Physics

c. 1600-1900
Galilei
Bacon
Descartes
Newton

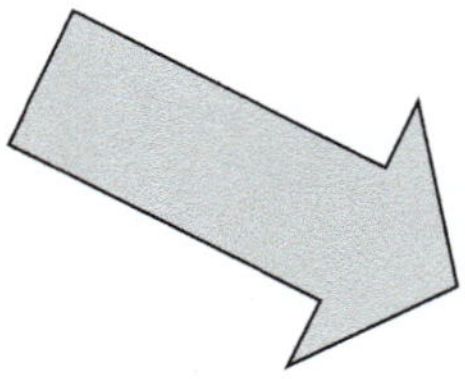

General Effects or Consequences:

End of Witch-hunt
Astrology excluded from science
The Enlightenment (18th century)
The New Enlightenment (19th century)
Much of Modernism (20th century)

Empiricism
Materialism, naturalism
Positivism

The ideal of scientific method
Belief in science, reason, empiricism, experiment, laws, determinism, certainty.
Confidence in reason.

The idea of progress
Secularization
Social engineering

....

Modern Biology

Darwinism, genetics

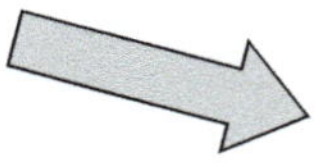

Physicalism, naturalism, historicism, empiricism, positivism, reductionism, secularism

....

Modern Physics

c. 1900, 1920 →
Einstein
Heisenberg
Schrödinger

?

Relativism?
Skepticism?
Uncertainty?
Increased dynamism

Postmodernism?
Increased belief in religion?
Increased critique of science?

3 General Introduction to the History of Science and to the Philosophy of Science

T. H. Brobjer, *Atlas of the History of Modern Science 1500-2020*, https://doi.org/10.1007/978-3-031-76627-5_3

What Is Science, and What Does the Word Science Mean?

The Meaning and Etymology of the Word "Science"

The simplest meaning of science is organized knowledge. However, one needs to go further and specify in what sense it is organized – it needs to be organized in a relevant sense (the organization of biological species would hardly be science if it was merely alphabetical) – and one needs to separate knowledge from false knowledge and belief.

Word:	*Episteme* →	*Scientia* →	*Science* →	Science	*Wissenschaft*
Language:	Greek	Latin	Old French	English (and modern French)	German
Meaning:	(Structured and certain) knowledge. The opposite of *doxa*, opinion, appearance.	Knowledge. The Latin translation of the Greek *episteme*.	Knowledge	A systematic approach to build and organize knowledge in the form of testable explanations (arrived at by observation and/or experiment through induction and/or deduction). Science is also used to refer to a body of knowledge that can be rationally explained and reliably applied.	Etymology: Goes back until at least 8th century old German, as a variant of *wissen* (*wissan*), something known or learnt. Until the 15th century, *Wissen(t)heit*. In the 18th century, the word *Wissenschaft* replaced *Kunst* as the word for a learned discipline.

The word scientist was coined by William Whewell in 1834, in parallel to the word artist.

Major Developments in the History of Scientific Thought

The two things which primarily characterize modern (post-1600) scientific thought compared to earlier times are the making of concrete experiments and the use of quantitative (rather than merely qualitative) arguments and methods, that is, the increased importance of mathematics (as a method, not as an *a priori* metaphysical principle).

Simplified overview of scientific thoughts on methods:

Period:	Antiquity	Middle Ages	The Scientific Revolution
Approach:	Observation	Authority	Observation, experiments and applied mathematics
Source (arguing for and/or representing this approach):	Aristotle	The Bible	Bacon, Galilei, Descartes and Newton

Somewhat more detailed overview of scientific thoughts on methods:

Period:	Greeks	Romans	The Middle Ages	The Renaissance	The Scientific Revolution
Approach:	Scientific observation (and a few experiments) with reasoning. Mathematics and deduction important	Little observation	Authority (of the Bible)	Authority (the Bible is exchanged for the ancients) and observation	Experiment, observation and mathematics
Characteristics:	Birth and early development of science	Little understanding of theoretical science	Faith and authority placed above science. (During especially 1400s, revival of Greek texts and science)	Revival of observation – partly due to the example of the Greeks	The beginning of systematic experimentation and use of quantitative methods

The Simplest Possible View of the History of Western Science 1

The Pre-Scientific Worldview

Viewed nature and cosmos as governed by magic, mysticism, religious and occult (hidden) forces.
Unclear distinction between matter/immaterial, life/non-life, human/animals/gods.
Almost always strongly anthropocentric.
Understands nature and cosmos in terms of purposes, reward and punishment.
No fact/value distinction (values are not separated from facts).
Organic, everything in nature and the cosmos is alive and related.

The Ancient Greeks, c. 600 BC – AD 200

Created and developed science and scientific methods and explicitly excluded religion from science, but also kept much of the worldview of the pre-scientific era: teleology, unclear distinction between material/spiritual, life/death, etc
Cosmos as rationally constructed. Greek culture and worldview thus contained many disparate elements, from (almost positivistic) science to religious mysticism.

Epicureans Lucretius	Aristotle Aristotelianism	Platonism Neoplatonism	Traditional Greek religion and worldview

The Romans, Middle Ages and the Renaissance, c. 200 BC – 1550

Most of the worldviews during this long period shared many aspects with that of the pre-scientific era, strongly coloured by Christianity, and also by Aristotelianism and Neoplatonic views.
Many occult forces or influences were assumed. Strongly anthropocentric.

The Mechanistic Worldview, c. 1600-1900

During this period, there was a strong orientation towards science, naturalism, reductionism, mechanism, atomism, physicalism, positivism, determinism, scientism, etc. Most explicitly in physics, but also in chemistry and biology, and even influencing the humanities.

The Modernist 20th Century, c. 1900- 2020

An enormous increase of science, scientific knowledge and technology – but also a (much?) humbler scientific worldview – there are limits to human understanding of nature and cosmos. The reductionism, positivism and determinism of the previous epoch have been significantly weakened. The natural world appears more complex than it did in the previous era.

The Simplest Possible View of the History of Western Science 2

Not Chronologically Proportional

-600

The Ancient Greeks

Created and developed science and scientific methods.

Hardly used experiments, overemphasized the deductive approach and sought after the purpose (*telos*) of everything.

200

The Roman Empire

The Romans hardly contributed at all to theoretical science. Important merely as transmittors.

Middle Ages

During the Middle Ages religion (faith) was placed above science. No or little progress made until towards the end (and then often through ancient Greek texts).

The Renaissance

The humanists exchanged religious dogma for the authority of the ancients. Mostly concerned with the arts, but eventually some scientific work, including interest in, and giving high status to, alchemy and astrology.

1600

The Scientific Revolution

Great expansion and many new paramount discoveries of the natural world. A continuation of the earlier Greek spirit, science and knowledge for its own sake, but now in opposition to Aristotle and teleological views.
Much more emphasis on the experiment, and the quantifying spirit (mathematization), which the Greeks theoretically argued for, now often became possible also in practice.

1700

The Enlightenment

1800

<-> Roman-ticism

19th Century Positivism

Great expansion of knowledge and of natural science

1900

The Modernist 20th Century

Enormous expansion of knowledge and of natural science. The advance in technology has been equally prominent.

Four Interpretations of How Modern, 20th Century, Science Relates to the History of Science: First Approximations

Not Chronologically Proportional

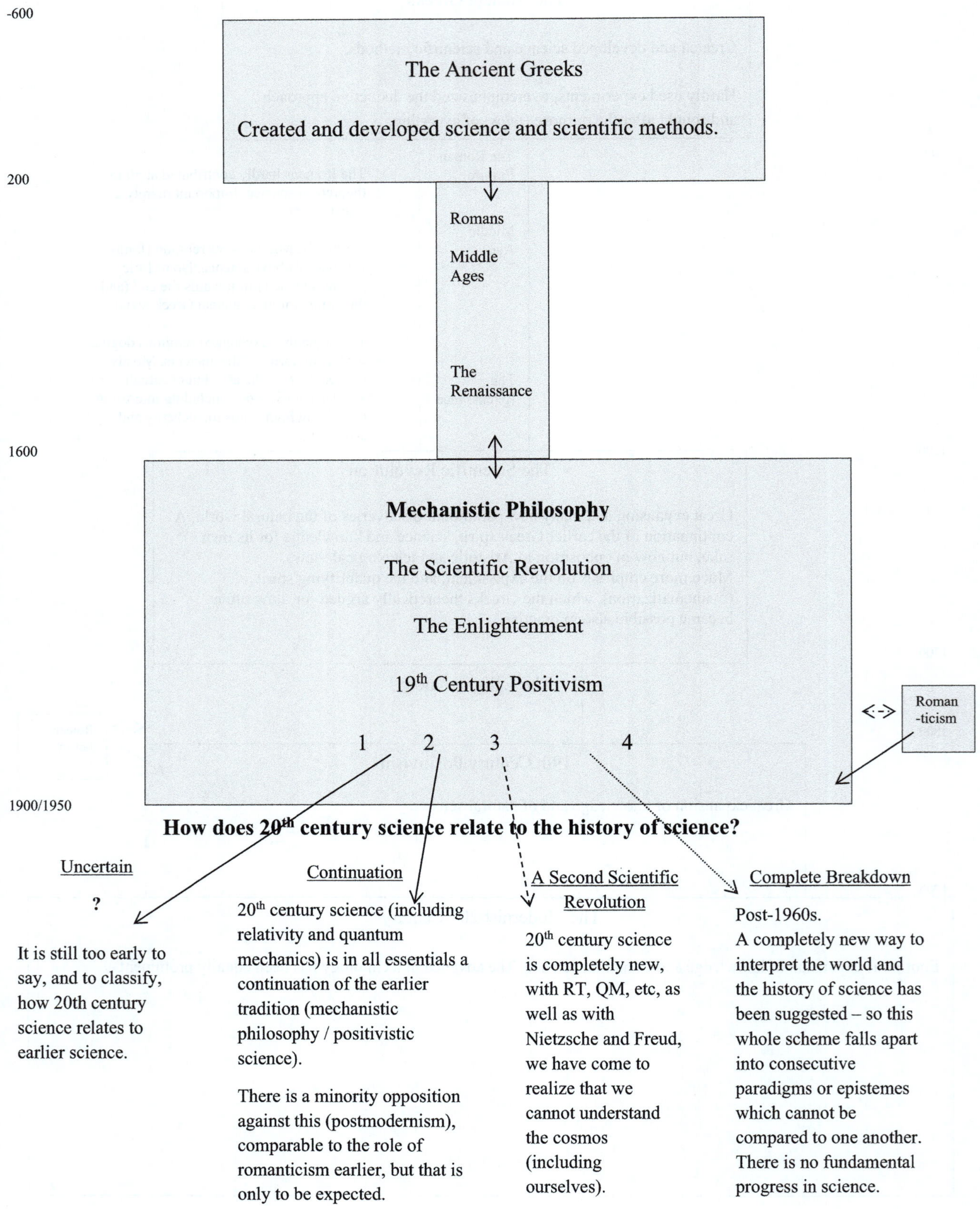

Philosophy of Science: Antiquity – 1700: 1. Overview

The Most Influential Thinkers Concerning Discussions of the Methods of Science

Not Chronologically Proportional

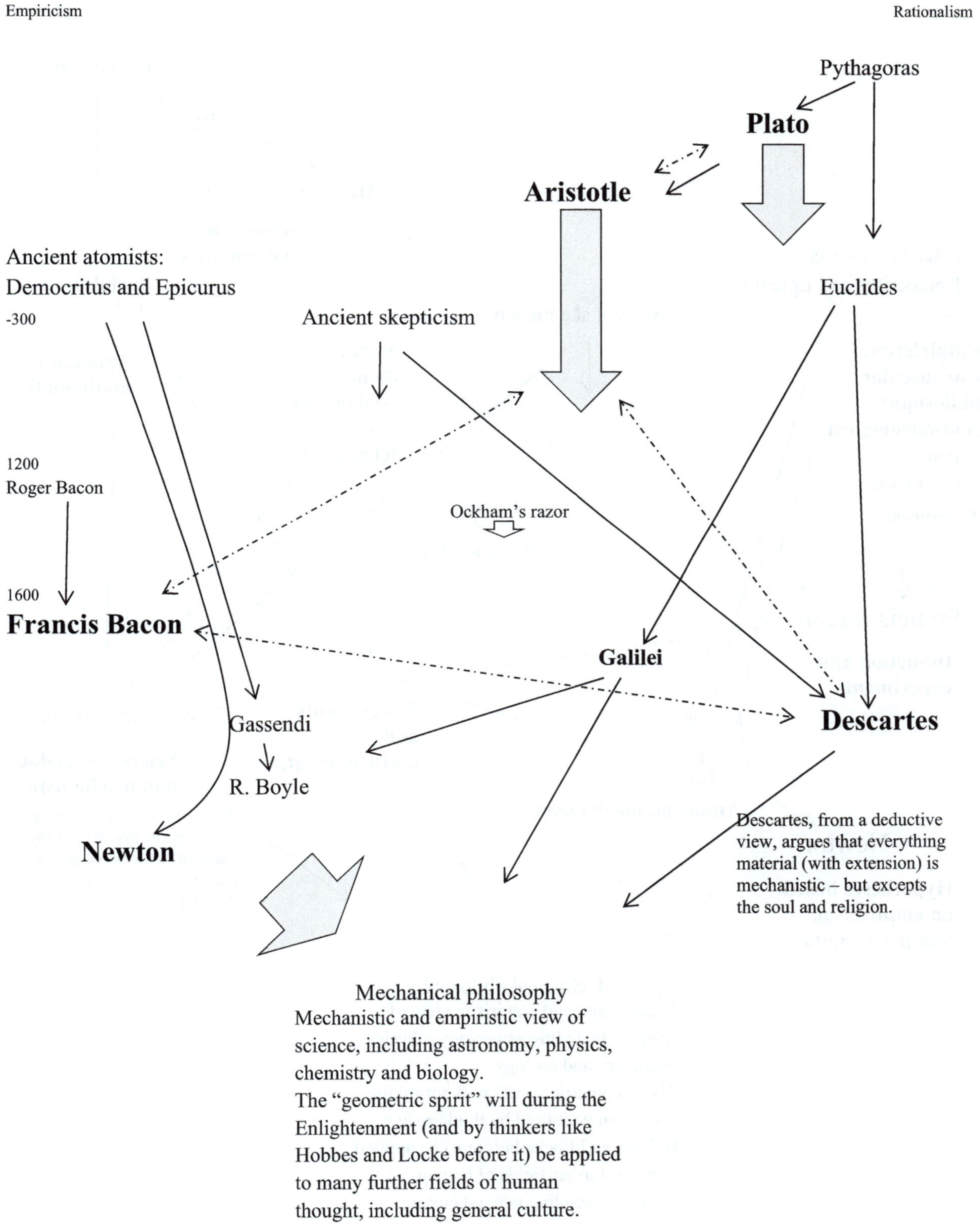

Philosophy of Science: Antiquity – 1700: 2. Overview

Principal Positions Regarding the Methods of Science

Not Chronologically Proportional

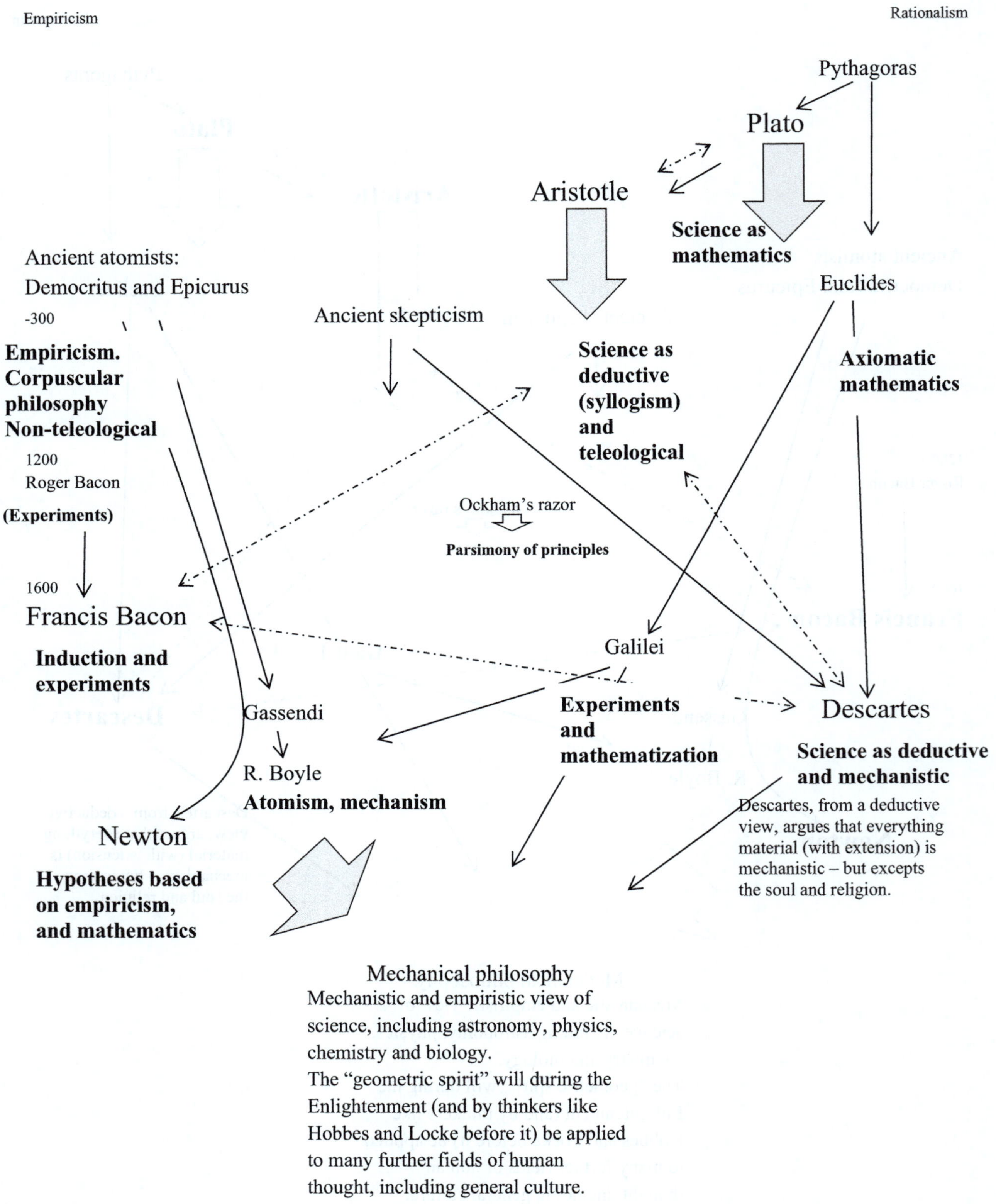

Philosophy of Science: Antiquity – 1700: Overview: 3. With Summaries

Discussions of the Methods of Science

Not Chronologically Proportional

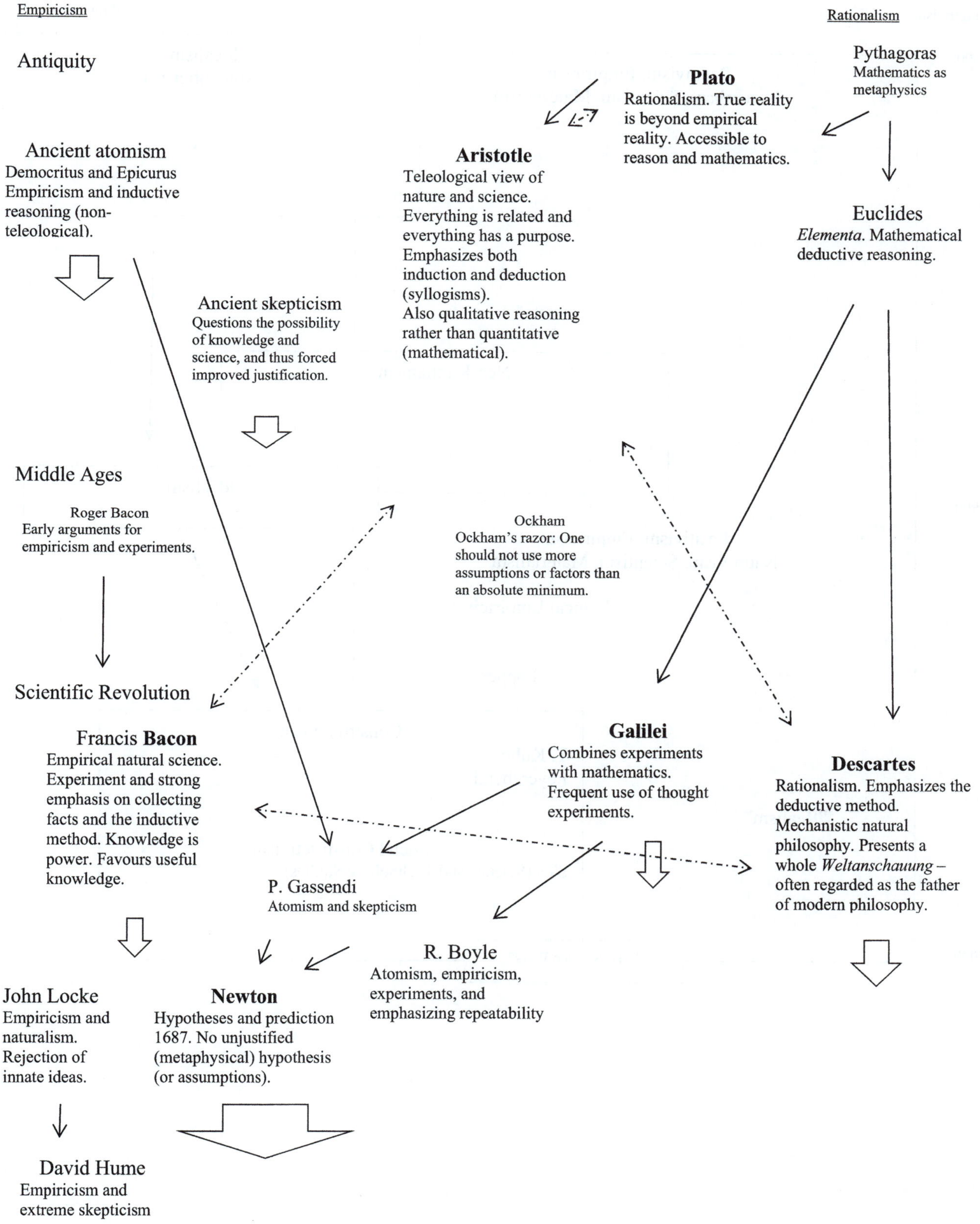

Philosophy of Science 1800–2000: Overview

"Scientific Philosophy"

Chronologically Proportional

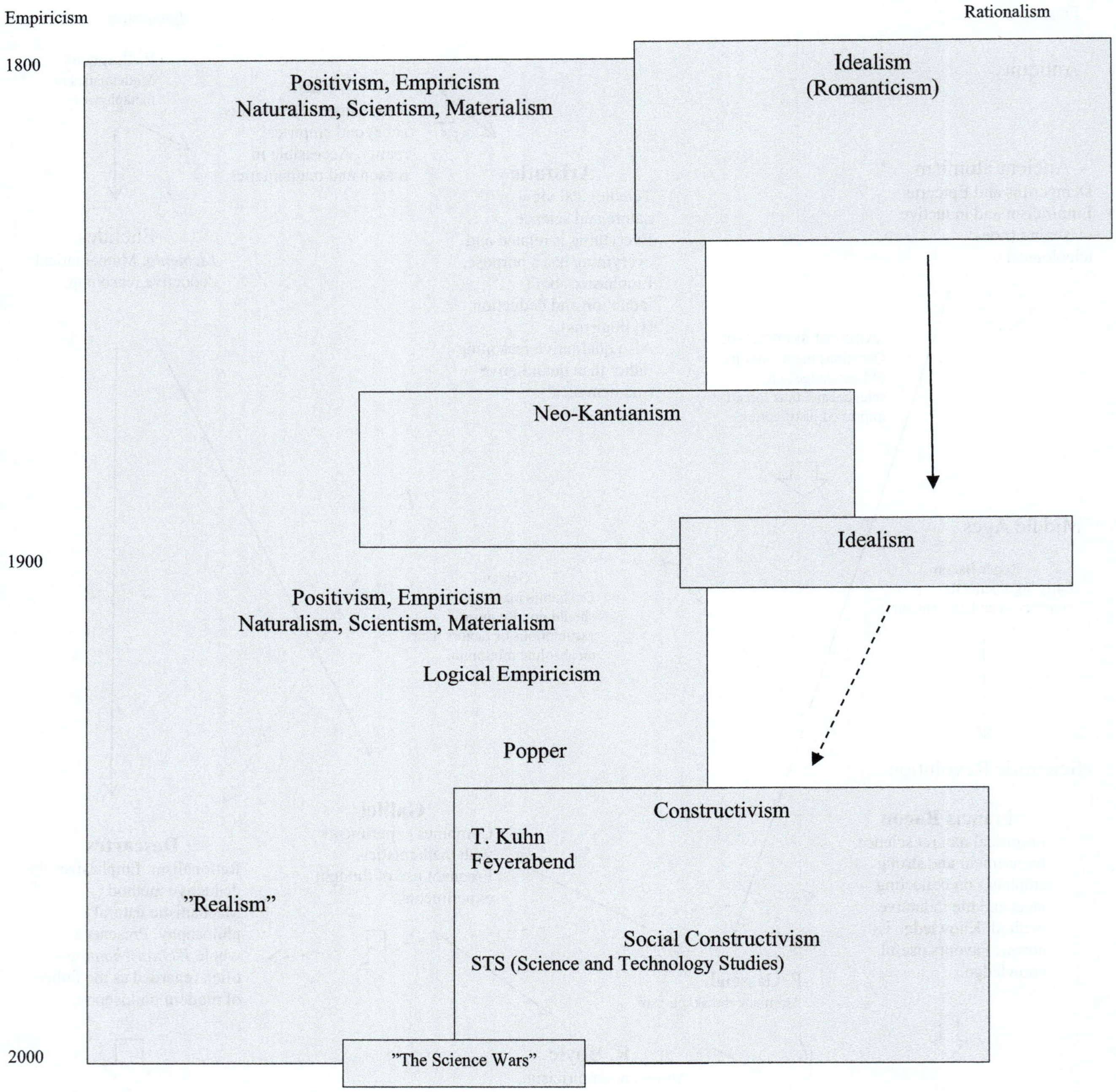

The History of the Experiment

Antiquity – 1900

Performing experiments holds a central place in modern science and in modern philosophy of science (although its importance has also been questioned since c. 1980). Although the Greeks made some experiments, these seem to have been few, and experimentation played little role in thinking about science until a beginning was made in the late Middle Ages. It was in the 17th century that experimentation was done systematically and given this central role both in scientific practice and the view of science.

(An interesting "opposite" variant of experiment is "thought experiment", *Gedankenexperiment*, which has played an important role in the history of Western science, see next chart.)

Empiricism and Induction
Observation and Experiment

Aristotle

He emphasized observation rather than experiment as the method to acquire certain knowledge, which should then be used deductively. An important purpose of science for Aristotle and most thinkers until c. 1600 was to find the teleological purpose behind everything that happens.

However, Straton, the third leader of the Lykeion made experiments in physics. As did several other Hellenistic thinkers and scientists, such as the physicans, Herophilus and Erasistratus, as well as Archimedes.

<u>The following thinkers argued for making experiments, and did some themselves</u>:

Grosseteste, c. 1250
Roger Bacon
Dons Scotus, c. 1300
W. Ockham

<u>With the thinkers listed below the experiment became established</u>:

William Gilbert, c. 1600
Francis Bacon, c. 1620

G. Galilei, c. 1620
I. Newton, c. 1687, 1704

<u>Other thinkers who have argued for the importance of experiments are</u>:

D. Hume
J. Herschel
J.S. Mill, c. 1850

↓

Rationalism
Deriving knowledge by deduction from self-evident first principles

Plato *et al*

Until c. 1600, the rationalistic approach was more emphasized and used than observation and the experiment.

Plato *et al* continued to be highly influential during and after the scientific revolution, due to the importance of mathematics – but now increasingly as a method of dealing with (observational or experimental) data – especially developed by Galilei – rather than as a first principle about the nature of reality.

↓

The History of the Hypothesis as a Method in Science (The Hypothetico-Deductive Method)

A hypothesis is a suggestion (not an axiom) taken as a basis for reasoning. The approach of using hypotheses lands somewhere between the experiment and the thought experiment, with predictions which are checked empirically closer to the former, without predictions closer to the latter.

Descartes, 1644

He emphasizes the importance of making hypotheses.

However, he and others realize that they do not lead to certain knowledge. Other solutions than the suggested explanation can be the correct one.

Newton, 1687

When Newton in the introduction to his *Principia* wrote that "I feign no hypotheses" this was aimed against Descartes' approach, and because he could not say and did not want to hypothetically suggest how gravity worked "at a distance".

This led to a rejection of or little explicit use of the method of using hypotheses for over hundred years, especially among empiricists and positivists – who generally instead argued for the inductive method.

For example, the hypothetico-deductive method was criticized in the introduction to the *Encyclopédie*, written by d'Alembert, 1751. Thereafter by J.S. Mill in the middle of the 19th century and by Ernst Mach at the end of it.

W. Whewell, in his *Philosophy of the Inductive Science* (1840) defended the importance of hypotheses, as did C. Bernard, c. 1865. They argued that the proper use of it should led to that *more* is discovered than was in the original hypothesis – and that the hypothesis should be used to make predictions.

During the 20th century there was an increased critique of the inductive approach, and the hypothetico-deductive became perhaps the standard view of how science should and does work (with a strong emphasis on making predictions that needed to be confirmed).

Popper, 1959, argued that one can never confirm and verify the predictions and the original hypotheses (other explanations than those considered will always be possible) – as had already previously been suggested by Whewell and Ch. Peirce – but when combined with falsification rather than verification, it is the best scientific method known.

The History of Modern Scientific Objectivity: 1. Outline

Objectivity is the idea that truth can be known independently of viewpoint, perspective or bias. Objectivity has a history, and the word itself, in its modern meaning, is fairly recent, from the 17th or 18th centuries, but some argue as late as mid nineteenth century. Objectivity is a complex concept, being closely related to and dependent on the view of truth, science, fact and reality, and further related to such concepts as subjectivism and relativism. This chart is based on the book *Objectivity* by L. Daston and P. Galison (2007), primarily based on the examination of visual scientific atlases, and the debate and reviews it engendered. The chart should be regarded as representing a debate more than the generally accepted results of such debates. One could also add a fourth, contemporary tradition, that attempts to avoid the use of objectivity in the study of the history of science.

Three forms of modern objectivity:

Forms of objectivity	True-to-nature	Mechanical objectivity	Trained judgement
Persona	Sage	Worker	Expert
Image	Reasoning	Mechanical	Interpretation
View of the mind	Passive and impressionable	Active (influenced by Kant)	Interpretative
Practice	Selection Synthesis	Automated transfer (e.g. the photograph)	Pattern recognition
Ontology	Universal (the underlying type)	Particulars	Families

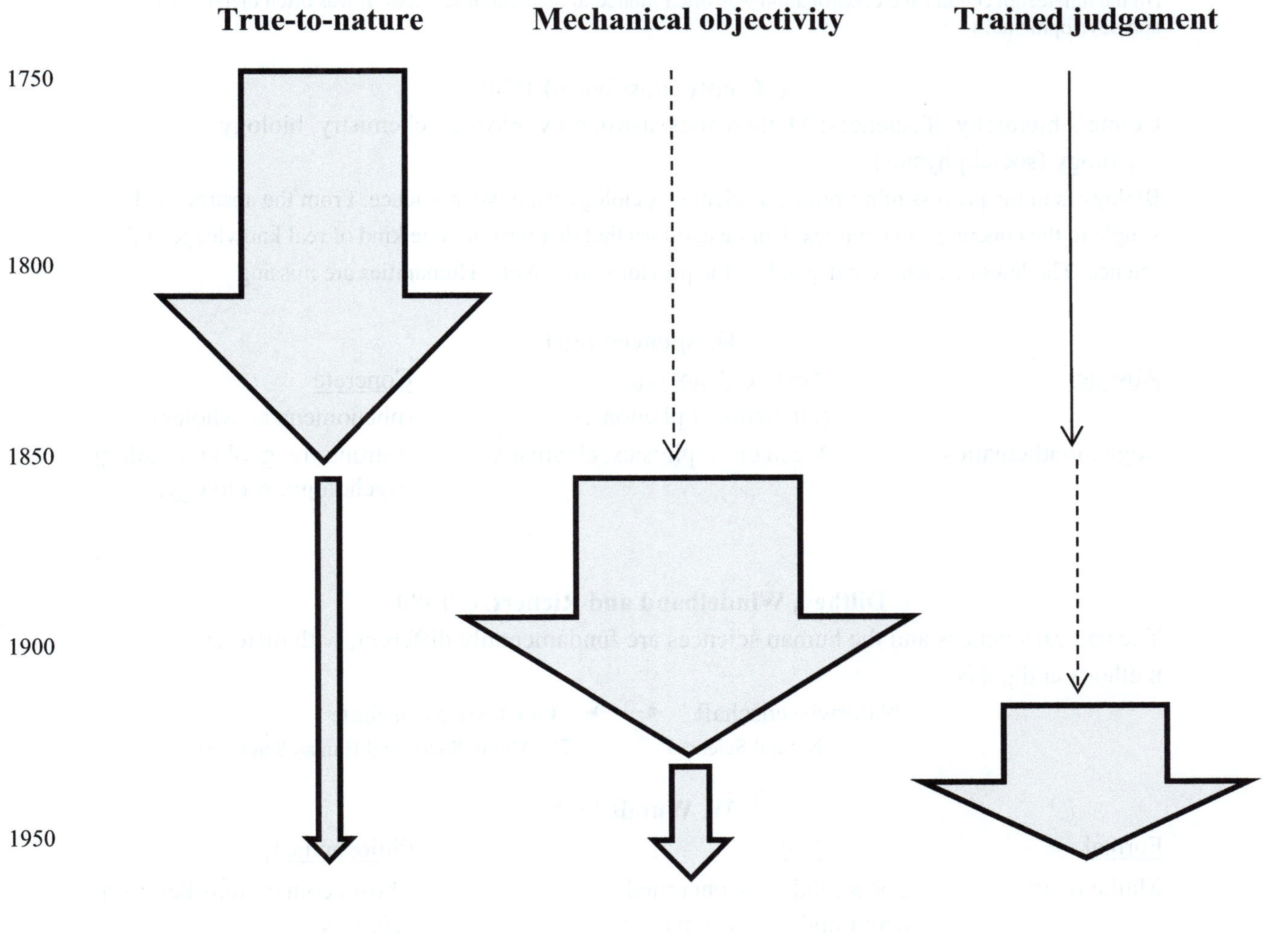

History of the Classification of the Sciences

Aristotle c. 330 B.C.

Theoretical	Practical	Productive
First philosophy, mathematics, physics	Ethics, (oeconomics) and politics	Productive and imitative
Objective truth	Practical truth	Not truth-related

Bacon c. 1620, classified according to human faculties

Memory	Imagination	Reason
History	Poetry	Philosophy
Natural and civil (human) history.		Divine, natural and human

⇩

Encyclopédie 1751

Almost identical to Bacon's classification, but less belief in that the classification was "natural". Mathematics included in the classification. Less theology and more practical science.

....

(In the nineteenth century the classification was often limited to the natural sciences. It was often hierarchical and developmental.)

A. Comte (Positivism) 1830

Comte's hierarchy of sciences: Mathematics, astronomy, physics, chemistry, biology, sociology (social physics).

Biology is in the process of becoming a science, sociology is not yet a science. From the abstract and simple to the concrete and complex. Comte assumes that there is only one kind of real knowledge and science. The laws of each field depend on the previous ones. Note: Humanities are missing

H. Spencer 1864

Abstract	Abstract-Concrete	Concrete
	(elements of phenomena)	(phenomena as wholes)
Logic, mathematics	Mechanics, physics, chemistry	Astronomy, geology, biology, psychology, sociology.

Dilthey, Windelband and Richert, c. 1880

The natural sciences and the human sciences are fundamentally different, with different methods and goals.

Naturwissenschaft ⟷ Geisteswissenschaft
(Natural Science) (The Moral, Social and Human Sciences)

W. Wundt 1889

Formal	Real		Philosophical
Mathematics	Concerned with mind	Concerned with nature	Shows connections between sciences

Historiography of Science 1850–2000: 1. Overview

Main Trends and Schools

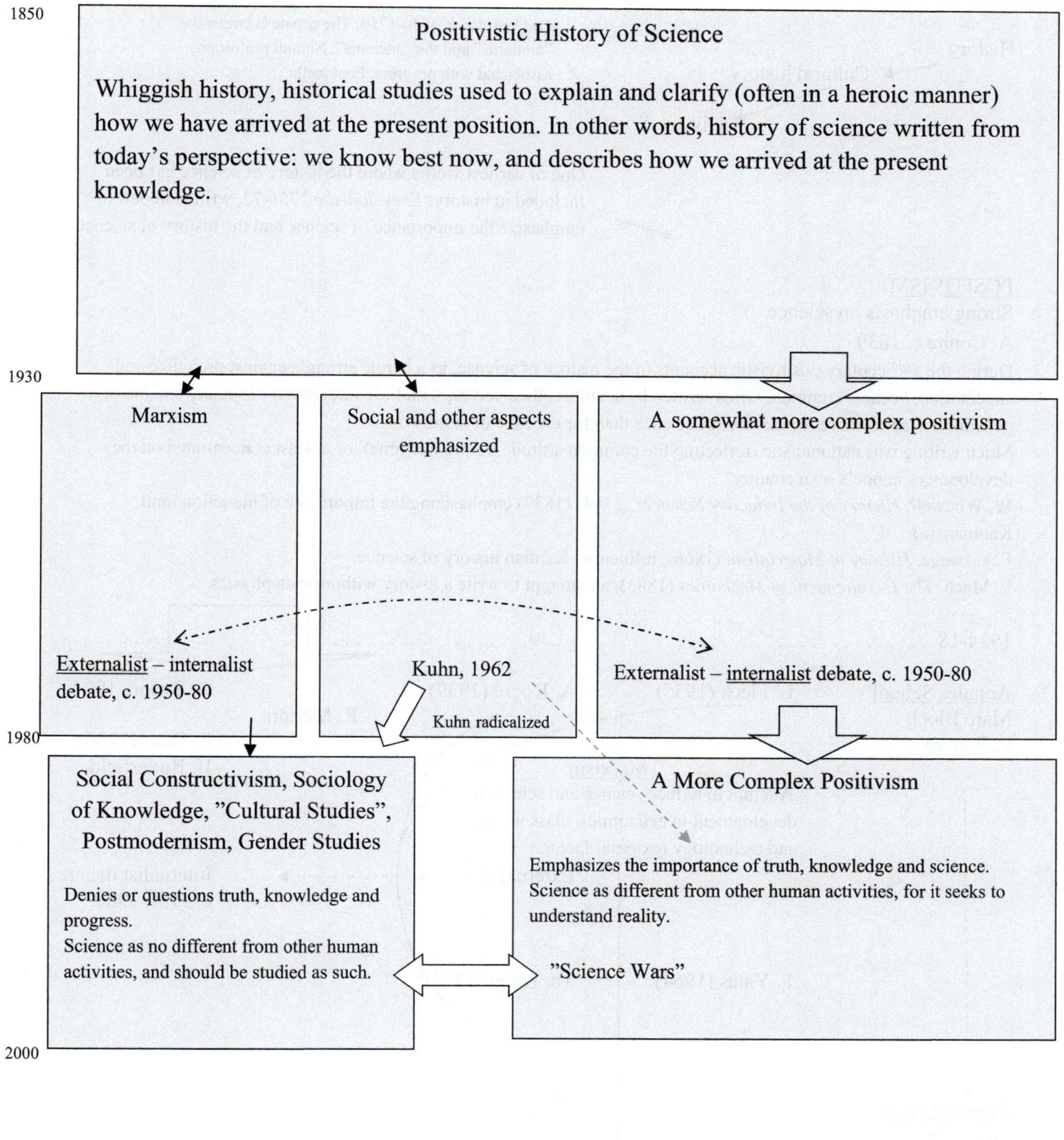

Historiography of Science 1700–2000: 1. Overview

How has the History of Science been Understood and Presented?

History → Cultural history → History of science

La Querelle, c. 1680-1710. The debate between the "moderns" and the "ancients". Natural philosophy associated with progress. Fontanelle.

One of earliest works where the history of science has been included in history: *Encyclopèdie* 1751-72, written in part to emphasize the importance of science and the history of science.

<u>POSITIVISM</u>

Strong emphasis on science

A. Comte c. 1830

During the 19th century positivistic accounts of the history of science, as a heroic struggle against prejudice and superstition, became common. Often written by scientists themselves, with little integration of society and history generally. Written more for the sake of science than for the sake of history.

Much writing was nationalistic (reflecting the common attitude during this time), or at least concentrated on the development in one's own country.

W. Whewell, *History of the Inductive Sciences*, 2 vols (1837) emphasizing the importance of induction (and Kantianism)

F.A. Lange, *History of Materialism* (1866), influential Kantian history of science.

E. Mach, *The Development of Mechanics* (1883), an attempt to write a history without metaphysics.

1914-18 ....

<u>Annales School</u>
Marc Bloch
L. Febvre

L. Fleck (1935)

A. Koyré (1939)

R. Merton

G. Sarton, 1927

E. Burtt (1950)
H. Butterfield

<u>Marxism</u>
Attempt to reduce science and scientific development to economics, class warfare and technology (external factors)

Externalist ←·-·-·→ Internalist debate c. 1950-1980

F. Yates (1964)

Th. Kuhn (1962)

<u>Postmodernism</u>
Foucauld

<u>Constructivism</u>
S. Schaffer
S. Shapin

Sociology of knowledge 1970→
Sociology of scientific knowledge (SSK)
STS. Barry Barnes and Bruno Latour

Historiography of Science 1900–2000: 2. Overview

How the History of Science has been Understood and Presented?

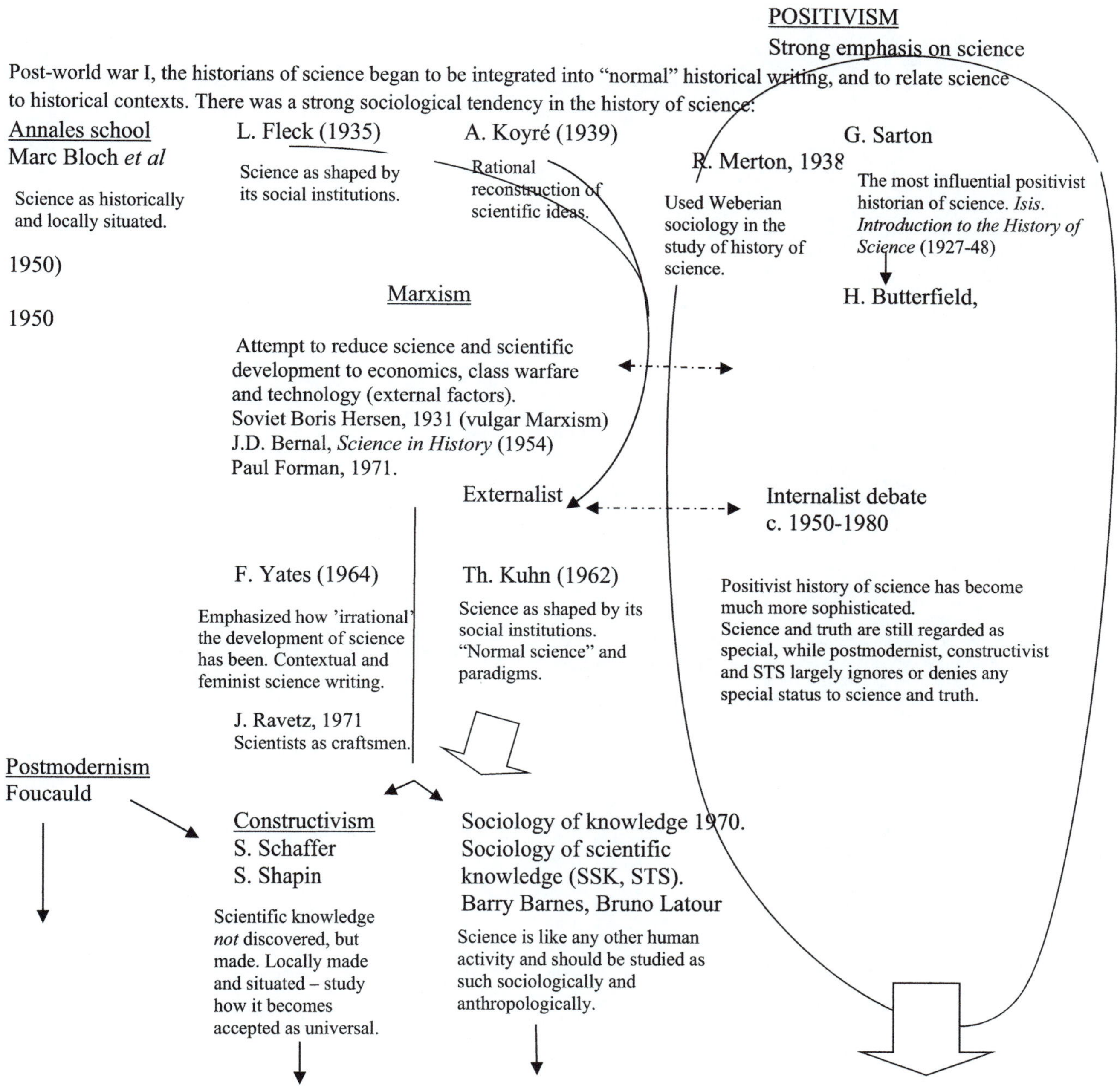

4 History of Physics

Many more specific and limited charts are presented in Chap. 6 below.

T. H. Brobjer, *Atlas of the History of Modern Science 1500-2020*, https://doi.org/10.1007/978-3-031-76627-5_4

The History and the Interrelation of the Major Fields of Physics (c. 1500–2000): 1. Overview

Mechanics, explaining how bodies move, was until 1900 the central and exemplary field of physics (and also beyond physics and science).

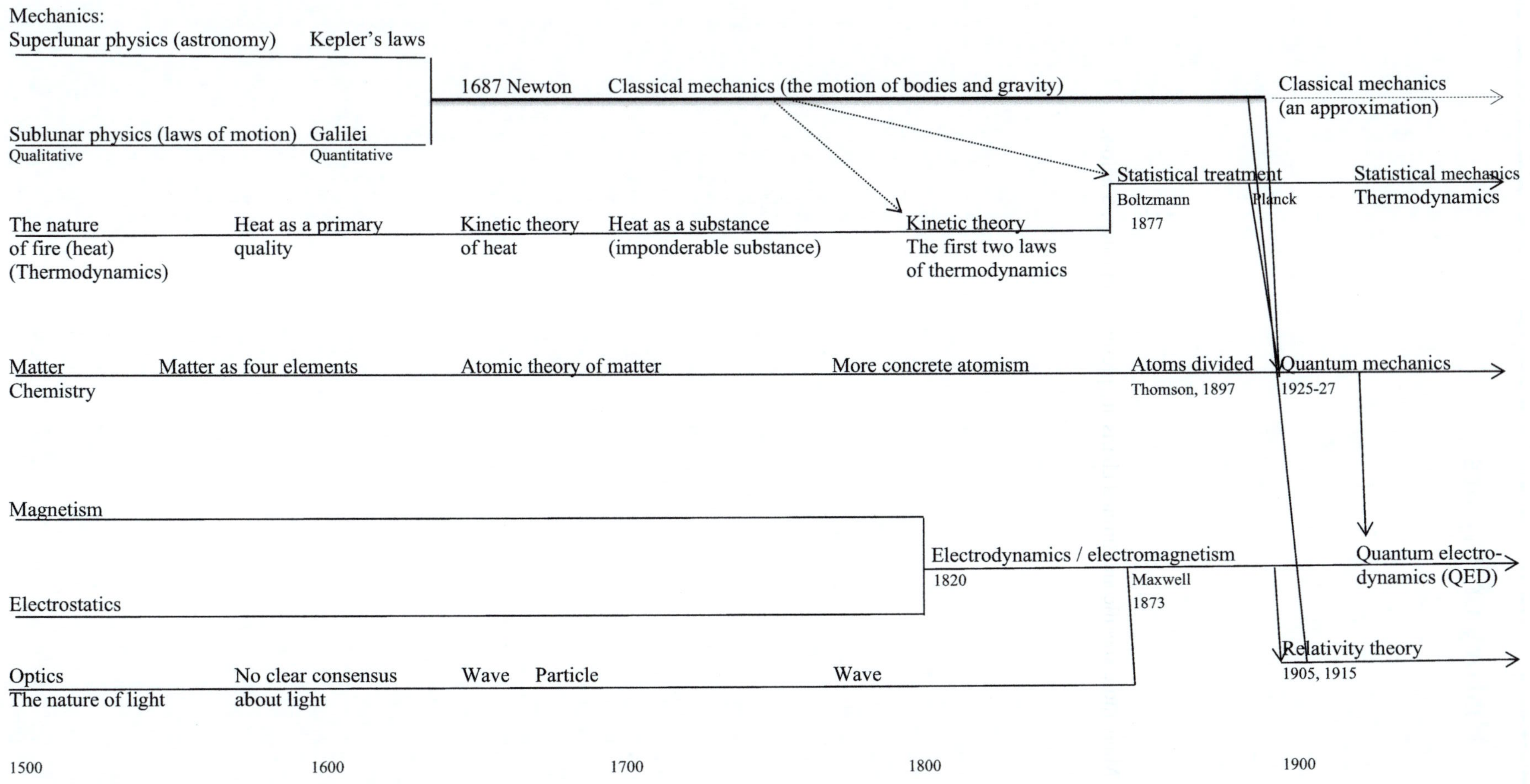

The History and the Interrelation of the Major Fields of Physics (c. 1500–2000): 2. With Names

With the Most Important Scientists

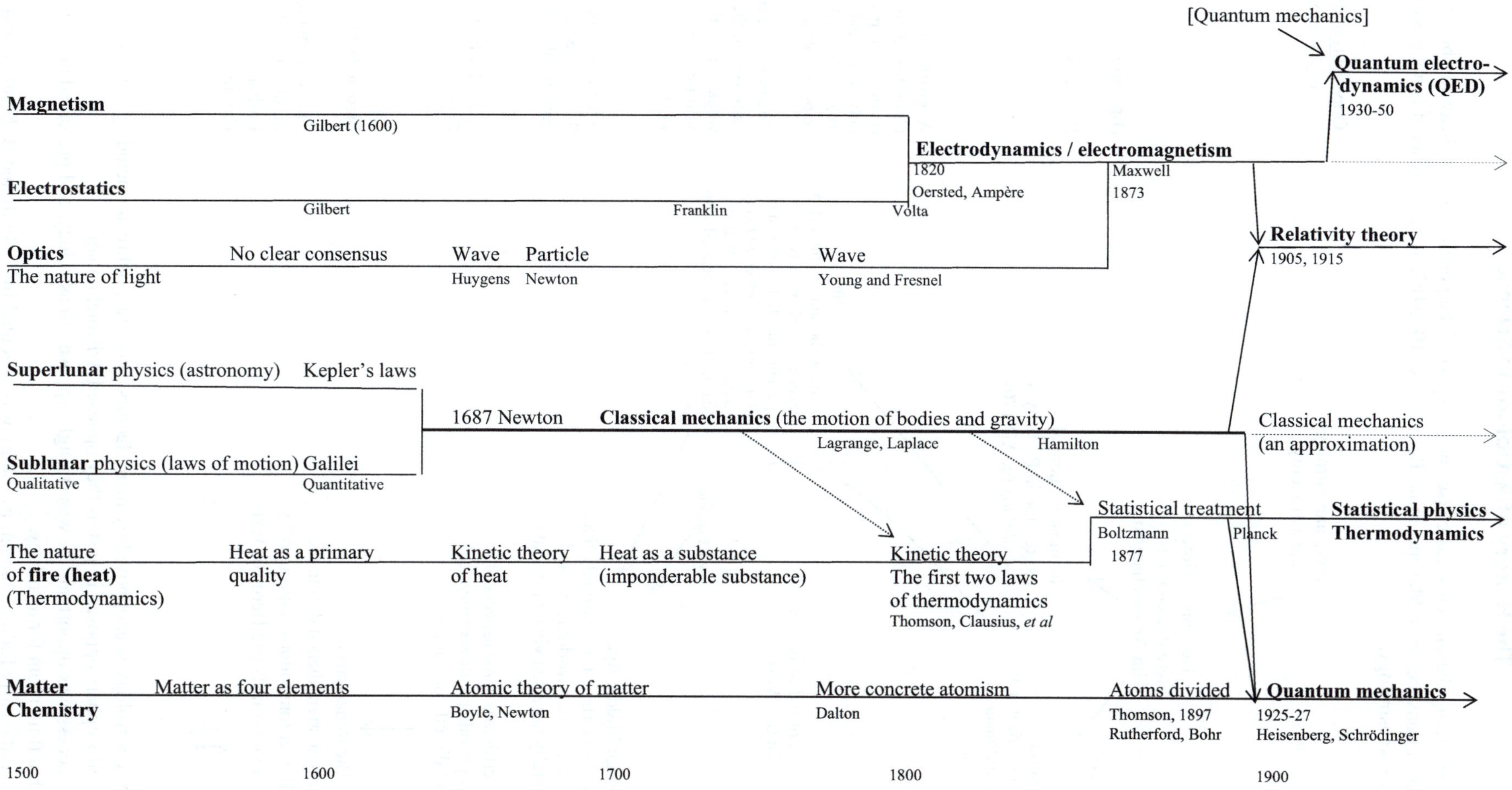

The History of Astrology in Science

Astrology, the theories concerning the influence (as causes or as signs) of the positions of the heavenly bodies in terms of human and terrestrial implications, began in Mesopotamia (Babylon), c. 1000 BC, at a time when it is not possible to distinguish between astronomy and astrology.

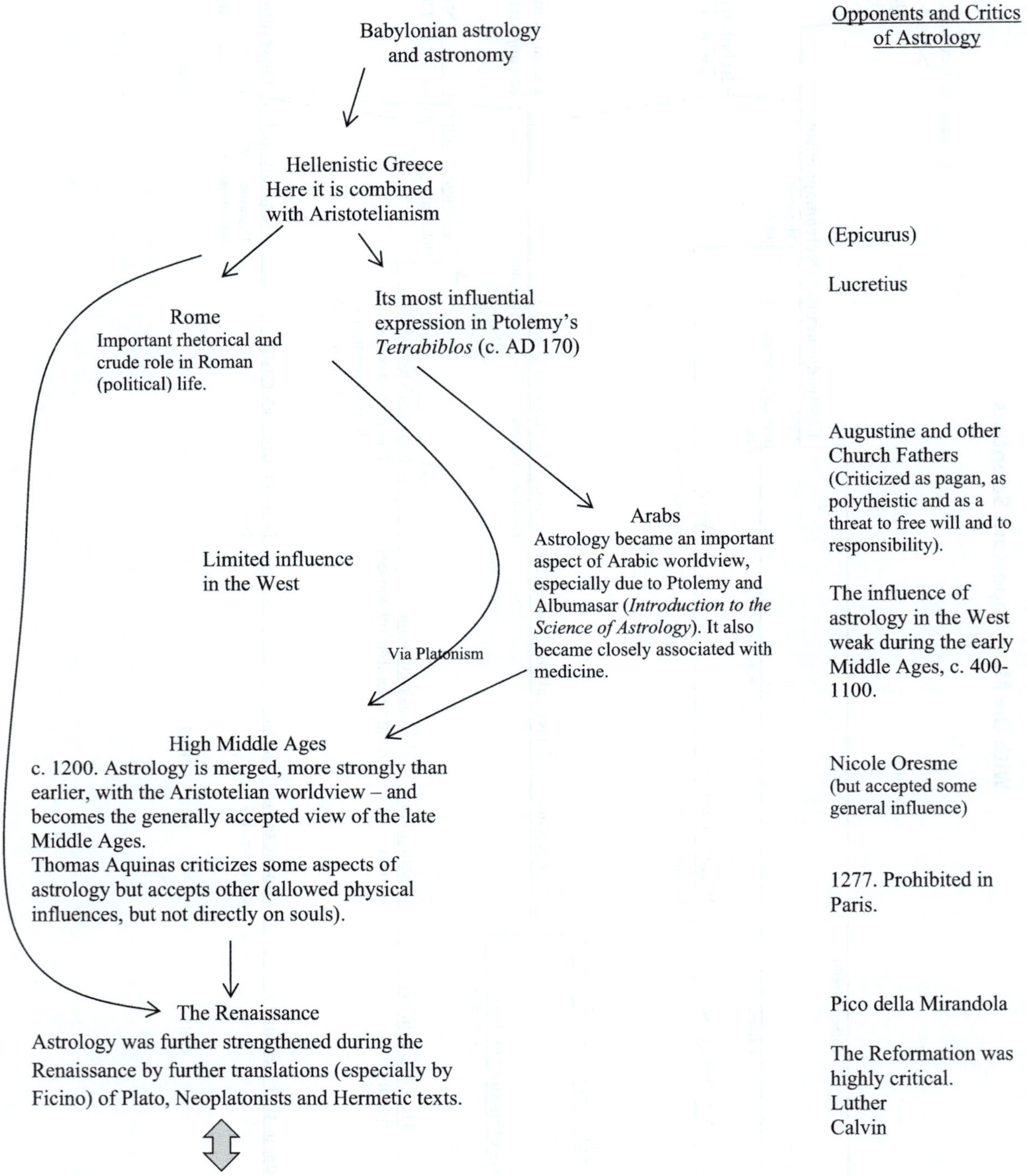

During the 17th century astrology was opposed by many factors, religious, political (including centralization of power) and most of the natural philosophers opposed it, just as they opposed Aristotelianism.

Especially Newton's success was important. He was strongly opposed to astrology, and presented another alternative – classical physics – which left no room for astrology.

By the 18th century and the early Enlightenment, astrology was ridiculed by J. Swift and B. Franklin and many others, and had by then been excluded from science and natural philosophy.

The History of Alchemy in Science 300–1700

Alchemy probably originated in Hellenized Egypt, in Alexandria.
The earliest known alchemical writer is Zosimus, c. AD 300, followed by several other anonymous writers during the next century. His and their texts are highly allegorical and allusive, which led to problems with interpretations.
The aims of alchemy:
1. The production of valuable metals (gold) from lesser metals (transmutation).
2. Pharmaceutical remedies
3. Philosophical or religious purification
Important in their worldview were the materials: Sulfur, mercury, arsenic and electrum (a gold-silver alloy). Especially sulfur and mercury more as principles than the elements known today.
Alchemy consisted from the very beginning basically of two parts:
1. Practical or empirical art, mostly aimed at transmutation.
2. A theoretical philosophy which were to explain or justify this (based on Aristotelian physics).

→ The Arabs, especially c. 900-1100

The Arabs came to emphasize and develop alchemy. The most famous Arab alchemist was Jabir ihn Mayyan (9th century) and the philosopher al-Razi (Rhazes), who wrote the influential *Book of the Secret of Secrets*.
Following early Greek alchemy, the Arabs emphasized that metals are formed by sulfur and mercury (normally in the depth of the earth), and thus by changing the proportions transmutation should be possible.

← c. 1100-1300

With the translation of new Geek texts and Arabic texts into Latin, c. 1100-1300, an increased interest in alchemy grew in Europe. For example, Albertus Magnus and Roger Bacon were interested in it. This interest culminated with **Geber**, c. 1300, who used strong mineral acids, distilled alcohol and other technological innovations at this time.
However, the interest in alchemy declined.

↓

The interest increased again during the Renaissance, c. 1450-1600, with new literary influences and new texts being translated, especially Platonic and the Hermetic texts. Alchemy was often combined with Christian mysticism, as well as with astrology (the fact that there were 7 known planets and 7 known metals increased this association).
Some alchemists worked primarily with practical tasks such as pharmaceutical preparations, techniques of smelting, assaying and other metal work or the production of dyes and other chemical substances. Others sought to produce gold, find the elixir of life, also referred to as the "philosopher's stone" (which would cure any illness and procure eternal life), and still others saw their activity primarily as ethical or religious. Many mixed all these elements.
The most important alchemist of this time was **Paracelsus**, c. 1530, a physician, who emphasized the use of mineral material, and argued for using chemicals in pharmacy and medicine (iatrochemistry).

↕

The influence of alchemy decreased greatly during the 17th century, but the words alchemy and chemistry (chymistry) were often used interchangeably, and late 17th century scientists, like Boyle and Newton, were active in alchemical pursuits.

The influential chemical textbook by Nicolas Lewey, 1680, is the first of a series that attacked alchemy as fraudulent – and by the early Enlightenment alchemy had been discredited and theoretically expelled from chemistry. Actually it continued to exist for almost another century.

The History of Astronomy Until 1700: Three Models of the Cosmos

Geocentric Concentric Spheres	Geocentric Epicycles and Excentric Circles	Heliocentric
Plato (sets the problem: mathematics and circle) **Eudoxus** (27 mathematical spheres) **Callipus** (33 mathematical spheres) **Aristotle** (56 *real*, i.e. material, spheres)	**Apollonius** (Epicycles and excentric cycles. Probably mathematical. Also much detailed and mathematical astronomical work) **Hipparchus** (Epicycles and excentric cycles. Probably mathematical. Also much detailed and astronomical work, including extensive starmap) **Ptolemy** (Epicycles, excentric cycles and equants. Primarily mathematical. Summarized ancient Greek astronomy)	**Pythagorians** (contemplated a non-geocentric universe) **Heraclides** (contemplated the earth rotating around its axis. Student in the Academy) **Aristarchos** (develops and presents an advanced heliocentric system, including showing that the sun is much larger than the earth. He also calculated the relative and absolute distances to the moon and the sun. Assumed that the stars were very far away. However, no important astronomer or thinker accepted his system) ↓ Seleucus

↓

Cosmology during the Middle Ages and for the Arabs

Cosmology during the Middle Ages and for the Arabs

Philoponos and Simplicius (c. 500)

John Scotus Erigena [Geoheliocentric system: Mercury and Venus around the sun, but everything around the earth]

Nicolaus of Oresme (c. 1360) ["If it was not for faith, one could believe that the earth rotates"]

Nicolas of Cusa [Denies that there can be a limit to the universe. Infinite. A spear-thrower]

Copernicus, 1543 (Platonist) Heliocentric
(Simpler than the Ptolemaic system, but still requiring epicycles and less accurate).

Tycho Brahe, 1590
Detailed and accurate observations.
Compromise geoheliocentric system.

Johannes Kepler, 1610
(Platonist). Denies that heavenly motion must be circular and constant, instead elliptical and non-constant.

Galileo Galilei, 1620
Supporter of the Copernican system.
Studied the sky with telescope and made many new observations.
Created mathematical physics, and the fields of dynamics and mechanics (how bodies move). However, his law of motions assumes circular motion to be natural (and therefore does not accept Kepler's results).

Isaac Newton, 1687
Presents his three laws of motion, and can from them accurately calculate motions both in the stellar and terrestrial regions. Sets the agenda for the next two hundred years.

Histories of the Theories of the Size of the Universe

Not Chronologically Proportional

Experimental or More Concrete or Philosophical / Theoretical Views of a Finite Universe

c. 450 BC. Anaxagoras claims that the sun is a large burning stone, at least as large as the whole of the Peloponnese. An observed and discovered meteorite later strengthened his claim.

c. 320 BC. Aristotle. A finite cosmos (with uncertain size), but also surrounded by vacuum (which possibly means that in some respects it may be regarded as infinite, but this was not Aristotle's or his followers' view). The earth as spherical.

c. 280 BC. Eratosthenes and others accurately measure the circumference of the earth (and thus also knew that it was spherical).

c. 200 BC. Aristarchos calculates the sizes and relative distances to the moon and the sun, and concludes that the sun is much larger than the earth. Since no parallax of the stars is visible, the distances to them must be very large.

c. 100 BC. Archimedes in his book *The Sand-Reconner* uses Aristarchos' cosmos as the largest known, approximately 4 light-years (ly) in radius (about the distance to the nearest star according to modern views).

c. 180. Ptolemy. Makes better calculation for the distance to the moon, but worse for the circumference of the earth (so his and the earth until c. 1500 was just a third of its true size).

During the Middle Ages it was assumed that all the stars lie on a sphere at the same distance from the earth.

17th century. Stellar parallax was attempted to be measured (e.g. by R. Hooke) but no positive results.

c. 1685, Newton, following J. Gregory, measured apparent brightness (and assumed the sun and Sirius to have the same absolute brightness), with the result that Sirius ought to be about 15 light-years from the earth. (Actually it is 8.6 ly away, and 25 times as luminous as the sun.)

c. 1750, J. Bradley, concluded, from the failure to measure parallax, that the closest stars must be at least c. 6 light-years distant (which is in good agreement with today's knowledge).

1838. Stellar parallax measured for the first time. The distance to some stars measured.

Cephaids discovered. A special type of variable star that could be used to measure distances.

"The great debate" about the size of our galaxy and the universe. c. 1910.

Hubble shows that there are other galaxies, and measures redshift (expanding universe).

The discovery of CMB in the 1960s, confirmed the Big Bang theory, and thus seemed to limit the size and age of the universe.

Inflation. If this theory is correct (it is accepted by the majority of cosmologists) it means that the universe can be enormously much larger than the visible universe (which maximally is about $60*10^9$ light-years, due to the speed of light, c. 13.8, and due to the expansion of the universe) – and infinite in the Aristotelian sense of ever increasing faster towards infinity than anything can empirically test.

A Theoretical or Philosophical View that the Universe is Infinite

(This view existed from at least 300 BC, but was probably always a distinctly minority view.)

Epicurus, c. 300 BC

Lucretius, c. 50 BC

Nicolas of Cusa, 1440

G. Bruno, 1600

From Kepler's time this view was weakened by Olber's paradox (which seemed to show that the universe could not be infinitely large with infinitely many stars).

However, viewing the universe as eternal and infinite continued as a alternative.

Map of the History of Physics (Natural Philosophy) 1500–2000: 1. Overview

Chronologically Proportional

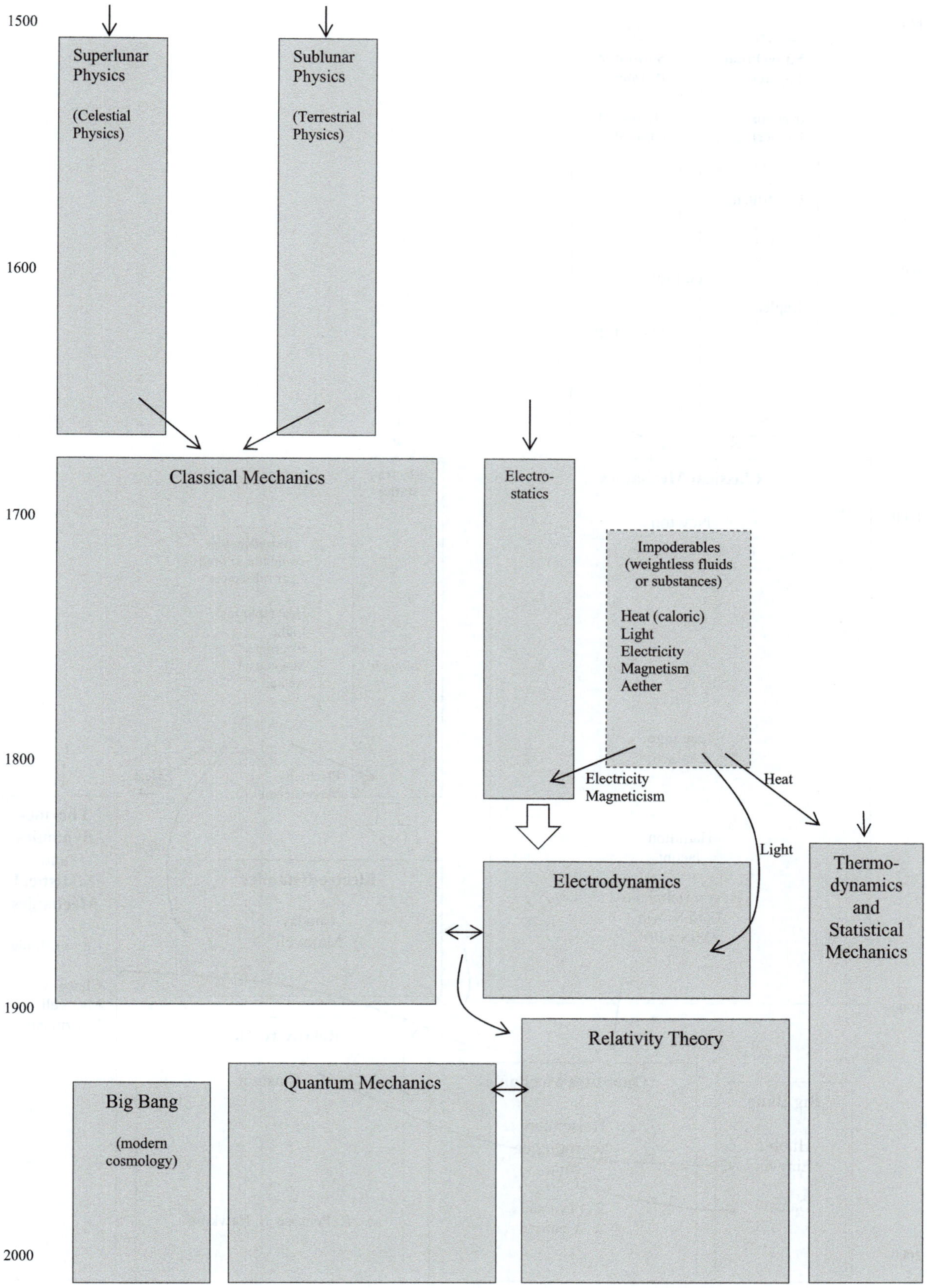

Map of the History of Physics (Natural Philosophy) 1500–2000: 2. With Names

Chronologically Proportional

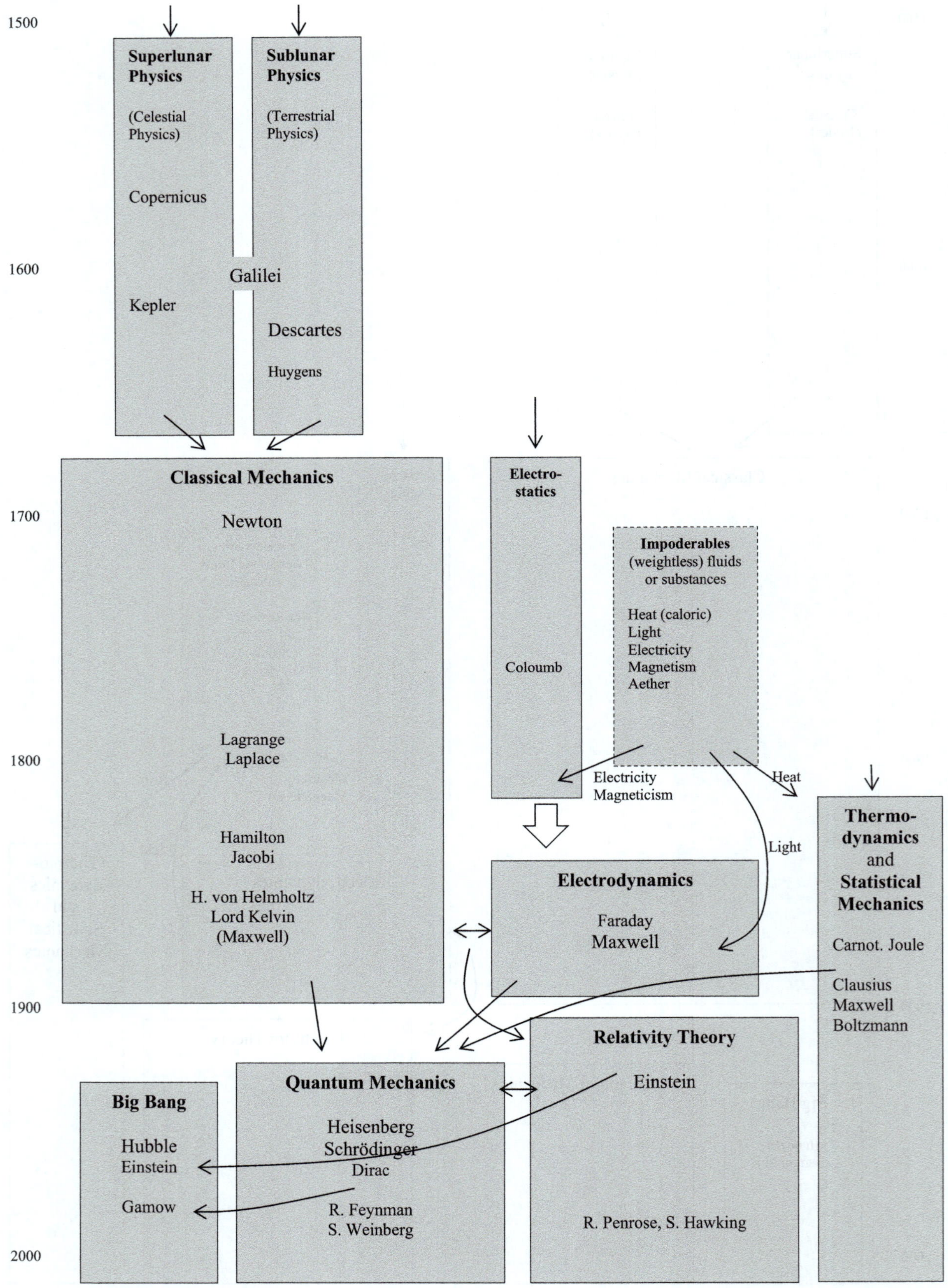

The Three Major Historical Kinds (Paradigms) of Physics: 1. Overview

Not Chronologically Proportional

330 BC

Aristotelian Physics

Teleological (everything has a purpose)
Motion problematic – still-standing natural
Objects seek their natural place
There is no vacuum
Geocentric
Qualitative
Four elements and/or three principles

c. 2 000 years

↕

1687

Classical Physics
(Newtonian Physics)

Empiricist
Non-teleological (instead mechanistic)
All motion is relative, therefore no difference between motion and still-standing
There exists vacuum
There exists absolute time and space
Heliocentric
Quantitative
Atomistic

c. 200-300 years

↓

1900

Modern Physics
Relativity and Quantum Mechanics

Empiricism
Non-teleological
All motion is relative – except the speed of light which is absolute – and in the atomic world we are unable to speak of motion (due to Heisenberg's uncertainty principle)
Vacuum is active, and has weight
Super-quantitative (we can calculate, but we are unable to conceptually understand many things)
Field-theories and elementary particle descriptions are mathematically equivalent

c. 100 years

The Three Major Historical Kinds (Paradigms) of Physics: 2. Overview

Chronologically Proportional

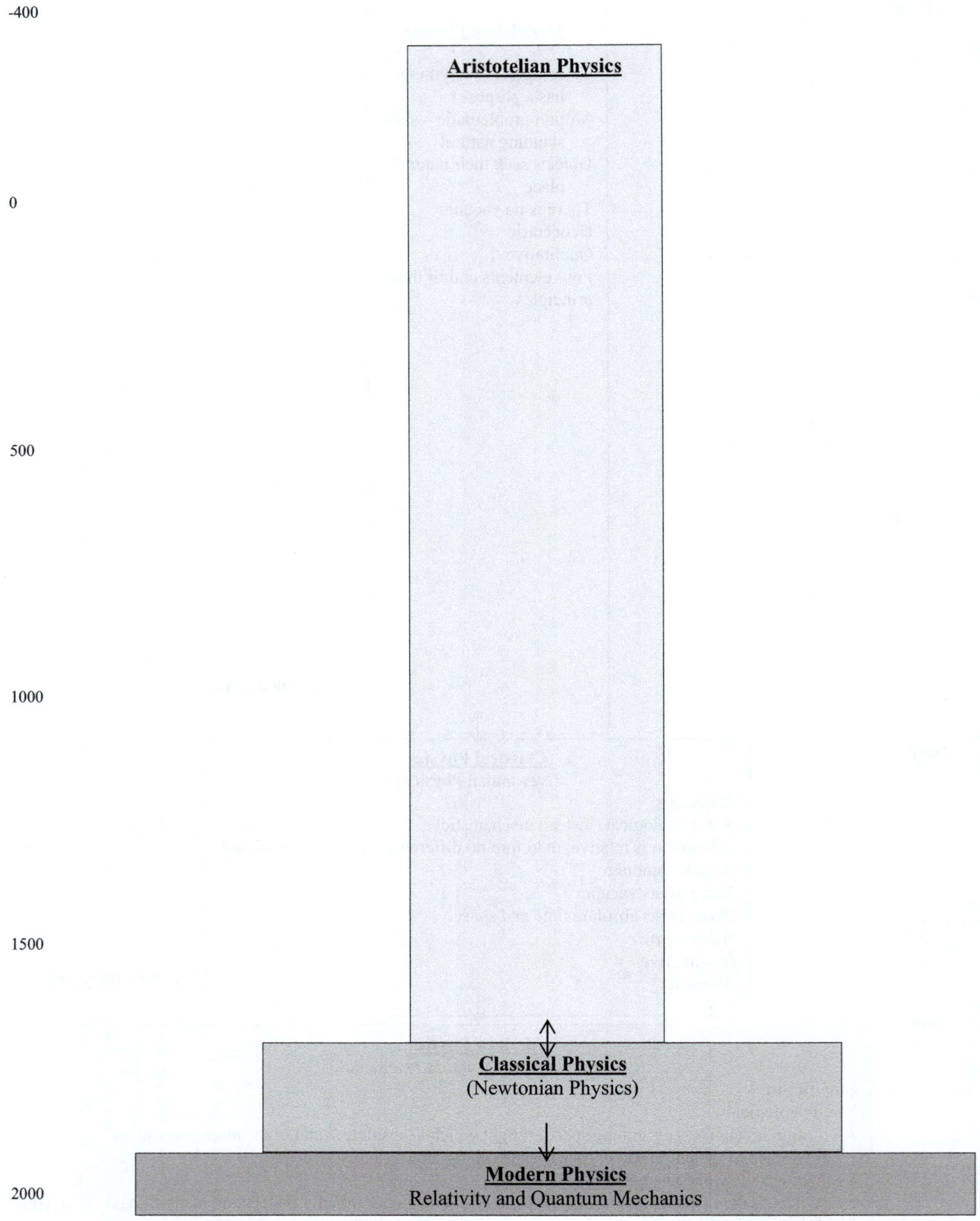

The History of Mechanics (Physics of Forces and Bodies) 1600–1900: 1. Outline

Mechanics (from Greek, *mechane*, machine), the study of effects of forces on bodies. After c. 1930 often referred to as classical mechanics.

Mechanics, with roots from Aristotle and Archimedes, was during this period the most central and influential part of physics.

Aristotle divided mechanics into two areas, in two separate ways:

Theoretical mechanics Natural movements	Practical mechanics Non-natural movements
Superlunar mechanics Celestial mechanics	Sublunar mechanics Terrestrial mechanics

↓ ↓

Copernicus
↓
Kepler

Galilei

Descartes
Huygens

Newton

Leibniz
Vis viva

Newton summarized and developed mechanics so completely that he determined the further development of the field for about two centuries. Although mechanics continued to be the central field in physics – and the prime example of ideal science also outside of physics, no outstanding new work was done and no outstanding other physicist could challenge his position.

Only relatively "minor" extensions of Newton's work was done in mechanics and hydrodynamics until c. 1900.

⇩

The great successes of mechanics, together with the law of energy conservation, made many in the second half of the nineteenth century believe that all natural phenomena might ultimately be represented and understood in terms of mechanics.

⇩

c. 1905 → Relativity Theory

c. 1927 → Quantum Mechanics

Quantum Mechanics
A new greater theory, including mechanics, also valid for small bodies and with a completely different conceptual view of reality and causation.

Relativity Theory
A new greater theory, including mechanics, also valid for fast moving bodies and with a completely different conceptual view of gravity (as well as of space and time).

The History of Mechanics 1687–1900: 2. More Details

Newton

Newton built his system and mathematics on three axioms or laws of motions:

(i) A body remains at rest or in constant velocity except when acted on by an external force (the law of inertia).

(ii) The force is equal to the mass times the acceleration (F=ma).

(iii) For every action, there is an opposite and equal re-action.

Although Newton almost "completed" (classical) mechanics, there were some further developments during the next two centuries. This occurred primarily in three areas; hydrodynamics (the behavior of fluids, liquids and gases) and related fields, new mathematical formulations of Newton's equations, and some aspects of statistical mechanics:

Hydrodynamics

J. Bernoulli
Hydrodynamica (1833), equations for ideal gases and liquids.

Maupertuis

D'Alembert, 1743
Equations of flow of inviscid (non-viscious) fluids and motion of rigid bodies.

L. Euler, 1755
The fundamental equations of hydrodynamics

c. 1820
Navier-Stokes
Equations for viscous fluids.

Cauchy
Equations for deformable solids

Alternative Mathematical Formulations of Newton's Equations

L. Euler
Extension of equations from applying to point-masses to rigid bodies

c. 1790. J.L. Lagrange
Mécanique analytique, 1788. New mathematical formulation, not in terms of forces but its integral over space, potential.

P. Laplace makes many new calculations and draws new conclusions in mechanics.

Poisson

c. 1830. W. Hamilton
New alternative formulation of the fundamental equations of mechanics, not in terms of forces, but in terms of energy.

↓

Jacobi
Simplified Hamilton's equations.

Statistical Aspects of Classical Mechanics

c. 1860
J. Maxwell
Statistical approach to systems of particles.

The History of Astronomy 1700–1900

More Theoretical, Often Affecting Worldview	Inbetween the Two Poles or Important for Both Aspects	More Observational
	Newton *Principia Mathematica*, 1687	Halley. The stars are not fixed (compared with ancient starmap).
		Halley. Predicted the return of comet, showing that it follows Newton's laws.
	L. Lagrange	1781. W. Herschel discovers Uranus.
W. Herschel makes detailed new stellar maps. Led to a three-dimensional view of star distribution (rather than the old view of all stars as approximately equidistant). The Milky Way as lens-shaped.	P. Laplace, *Méchanique celeste*, 1799-1825 Nebular hypothesis of planet and solar-system development. Allowed for a naturalistic developmental view of suns and solar systems.	1801. The first of the asteroids, Ceres, discovered. Led to an intense interest in the distribution of the planets from the sun (cf. Hegel's doctoral dissertation).
1838. The first measurement of stellar parallax, by F. Bessel, giving for the first time the distance to a star (c. 10 light-years).	Double-stars discovered, and could be shown to follow Newton's laws. c. 1800. Herschel discovers infrared (heat) radiation, beyond the visible spectra.	1845. The first of the great telescopes (180 cm mirror) finished.
	1846. Discovery of Neptune, using Newtonian calculations on the disturbances of Uranus.	c. 1845. Astronomical photography (but it will take until c. 1870 before films are sensitive enough to photograph stars).
1858. Spectral analysis of the sun. Helium discovered 1868. First spectrum of a star 1872 (due to better photographic plates). Led to new information about star composition, evolution and movements (through redshifts).		
1923. E. Hubble shows that the "Andromeda Nebula" lies far beyond our galaxy.		
1929. Hubble discovers that almost all galaxies are red-shifted, moving away from us, leading to the Big Bang theory (see below).		

The History of Cosmology 1600–2000

The main three cosmologies, with special emphasis on the Newtonian cosmology during the period 1687-1930, and some of the developments within it. Briefly listed are also the main earlier cosmologies, and the two main contenders in the twentieth century (see charts below for more on the Big Bang).

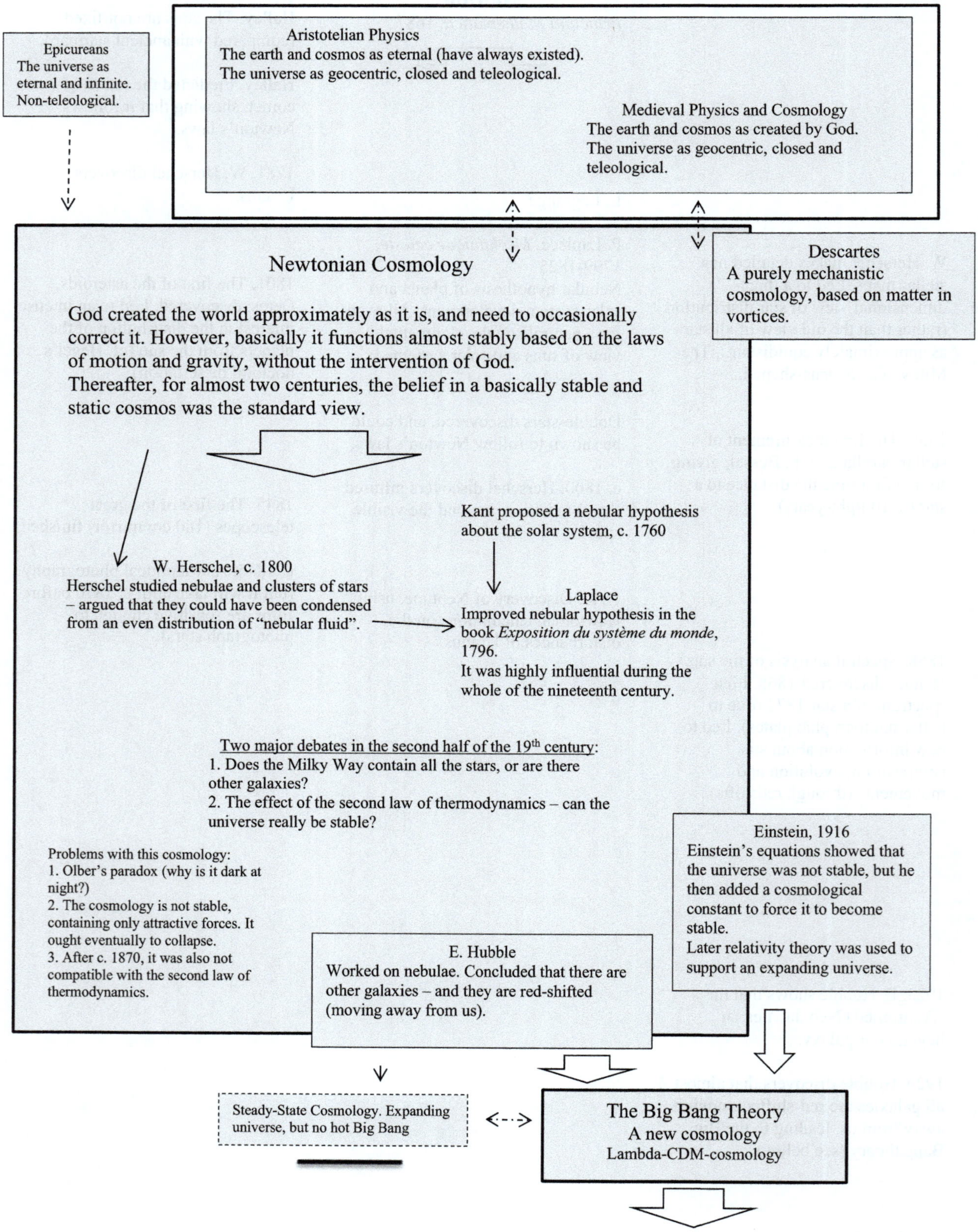

History of the Main Scientific Views of Matter 1500–2020: Overview

Chronologically Proportional

What does everything around us consist of? Are there many, few or only one single substance, and what does it or they consist of?

Almost all thinkers until circa 1800 (and many thereafter) assumed a dualism, a spirit/matter dualism, that assumed that apart from matter there also existed different versions of spirit, including God and the human mind and immortal souls, as acting on matter. This belief in dualism, mind/matter, is not included in this chart.

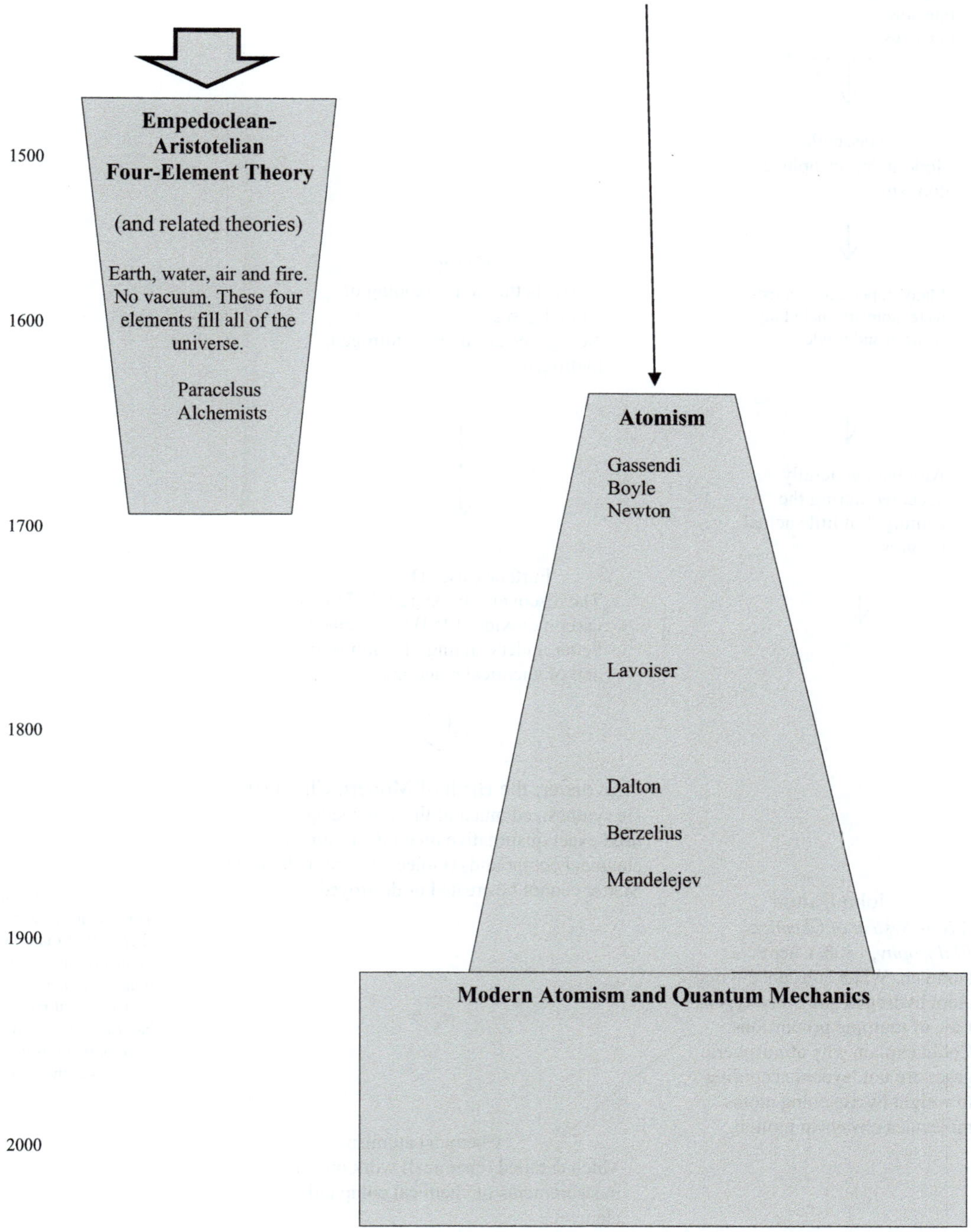

The History of Atomism: 1600–1900

During the period 1600-1850 atomism was primarily a philosophical and background scientific assumption that was accepted by ever more scientists, but with little empirical support and consequences, and some did not accept it.

More Theoretical or Speculative | More Experimental or Observational | More Speculative

1600

Ancient Atomism
Democritus
Epicurus
Lucretius

↓

Gassendi
Modern "philosophical" atomism

↓

Many important scientists were atomists, including Newton and Boyle.

Gases
A new better understanding of air and gases.
New gases discovered: nitrogen, hydrogen.

1700

↓

Atomism generally accepted during the century, but little actual progress.

↓

Further Gases Discovered:
The discovery of oxygen, 1774 (and carbon dioxide, 1754) led to much better understanding of combustion and of chemical reactions.

⇩

Lavoisier: the Birth of Modern Chemistry
He synthesized much of this knowledge, and used exact quantitative methods. Distinguished chemical compounds (molecules) from elements. Matter cannot be created or destroyed.

1800

John Dalton
A New System of Chemical Philosophy, 1808. Chemical atomism. Water substance = one atom hydrogen and one oxygen. Law of multiple proportions. Could explain why atmospheric gases are not layered according to weight by assuming atoms / molecules always in motion.

Romantic Science
The romantics were often skeptical of materialism and atomism. Instead they argued for polar forces, and occasionally for immaterial atomism – but this had mostly limited effect on the development of the exact sciences at the time.

⇩

Chemical atomism
Much detailed (chemical) work on exact measurements of chemical compunds and elements.
Berzelius. Atomic weights and symbols.
Avogadro.

1900

The Relation Between Classical Mechanics, Relativity Theory and Quantum Mechanics

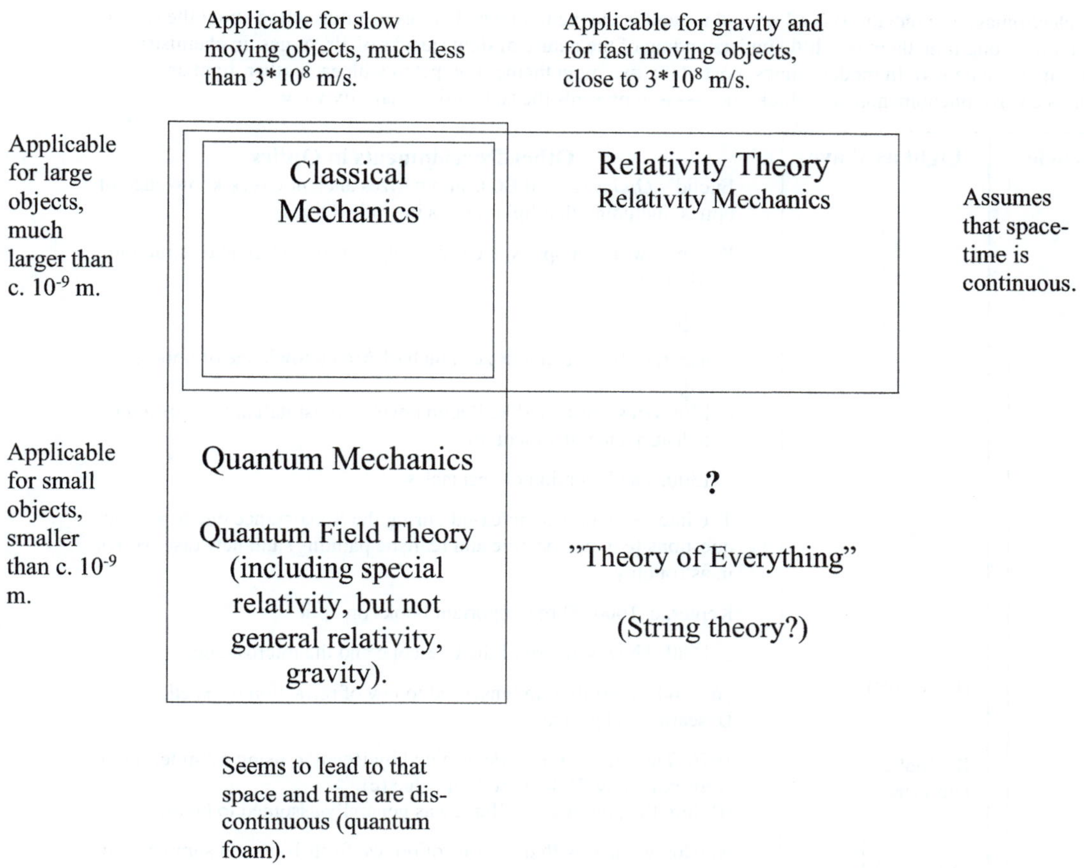

For us, in our "normal world", we have agreed to use the units of meter for length, kg for mass and seconds for time. These are human based units. Are there any natural units? The answer is yes, as Max Planck realized near 1900. The units of space, time and mass can be derived from fundamental constants and are called Planck length, time and mass (and are often used by physicists working in appropriate areas).

$l_{Planck} = \sqrt{\hbar * G_N \div c^3} \approx 10^{-35}$ m	G_N is the constant of gravity
$t_{Planck} = \sqrt{\hbar * G_N \div c^5} \approx 10^{-43}$ s	$\hbar$ is Planck's constant divided by 2π
$m_{Planck} = \sqrt{\hbar * c \div G_N} \approx 10^{-8}$ kg	c is the speed of light

The History of Optics and of Views of the Nature of Light

Light (visible electromagnetic radiation) is all around us, and is the medium which gives us information about the world through vision. For a long time there was little understanding of the nature of light (just as of air, or gas, in chemistry and of the zero in mathematics). In modern times, post 1600, two main themes competed with each other, light as corpuscular or as a wave phenomenon. The thick line, ▬, represents the standard or majority view.

Light as Particle	Light as Wave
Epicurus Lucretius	?
↓	(Descartes)
↓	R. Hooke Huygens
Newton (1704) ▌	(B. Franklin) (L. Euler) ↓
	Th. Young (1801) A. Fresnel ▌
	Maxwell ▌
	H. Hertz ▌

↘ Einstein, 1905 ↙

Explains the photo-electric effect and dismissed the existence of aether. Light has the character of *both* particle and wave – the light-quantum, later named photon (but it would take several decades before this was fully accepted). This was an important step on the way to quantum mechanics (QM) and QED.

The speed of light as constant (and the maximum speed) constituted one of the two axioms of relativity theory with great consequences for modern physics.

Other Developments in Optics

Euclid's *Optics* (c. 280 BC), summarizes ancient Greek knowledge of optics, including that light travels in straight lines.

Ptolemy, work on optics, including angle of reflection and refraction, c. AD 180.

↓

Alhazar, c. 1010, summarizes much of Arab knowledge of optics.

↓

c. 1200 Grosseteste and R. Bacon develop Aristotelian view of light as a disturbance of a medium.

c. 1300. The invention of spectacles.

The interest in optics increased during the Renaissance due to artistic questions (e.g. perspective and realistic painting) and new discoveries in astronomy.

Kepler, c. 1600. Three important books on optics.

c. 1600. The discovery of the telescope and the microscope.

1621-60. Aberration in lenses led to law of refraction by Snell, Descartes and Fermat.

1676. The speed of light determined by Roemer, using a Jupiter moon. Approximately 75% of modern value (c). (Before this, the speed of light was generally assumed to be infinite.)

Newton worked with questions of optics, from 1666 and summarized it in his immensely influential *Opticks* 1704. Light as corpuscular and that white light not "pure" but a mixture of all colours.

1800. W. Herschel discovers infrared radiation.
1801. J. Ritter, inspired by Herschel, discovers "invisible rays", i.e. ultraviolet radiation.

1801. Th. Young presents wave interpretation of interference.

1808. Malus discovers polarization of light – eventually a strong argument for the wave-theory.

1810. Goethe's *Farbenlehre*, directly aimed as a critique of Newton.

1817 Fresnel gives a mathematical formulation of interference and optics based on wave-theory – thereafter it begins to be the standard view.

For the rest of the century, many questions and problems with aether as the medium of light.

c. 1830. The begining of photography.

1850. Foucauld measured the speed of light as faster in air than in water and glas – supporting the wave-theory.

1878. Michelson measures the modern value of c.

1895. The first film shown by the brothers Lumière.

The History of Electricity (Electrostatics, Electrodynamics and Electromagnetism): 1. Overview

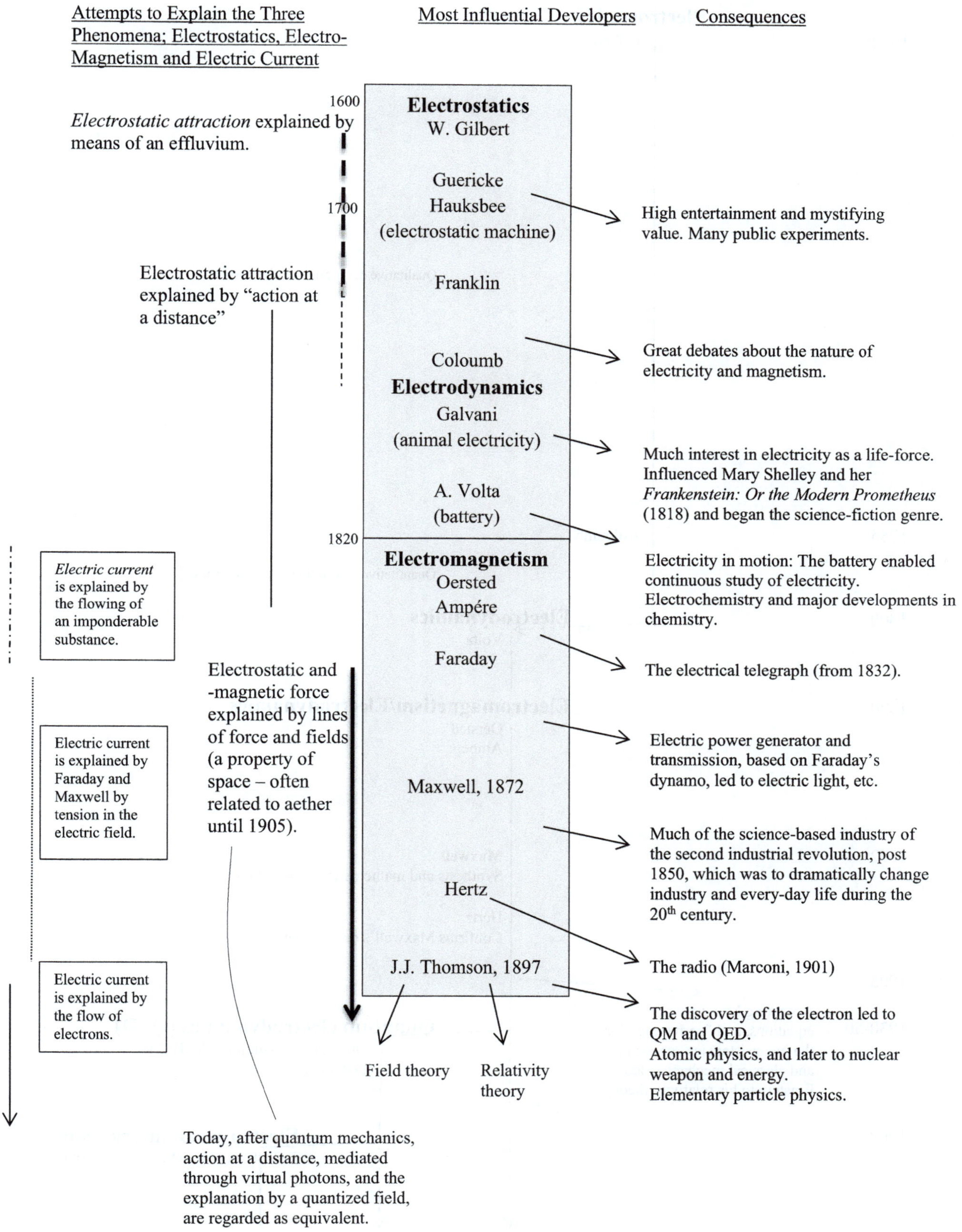

The History of Electricity: 2. Overview

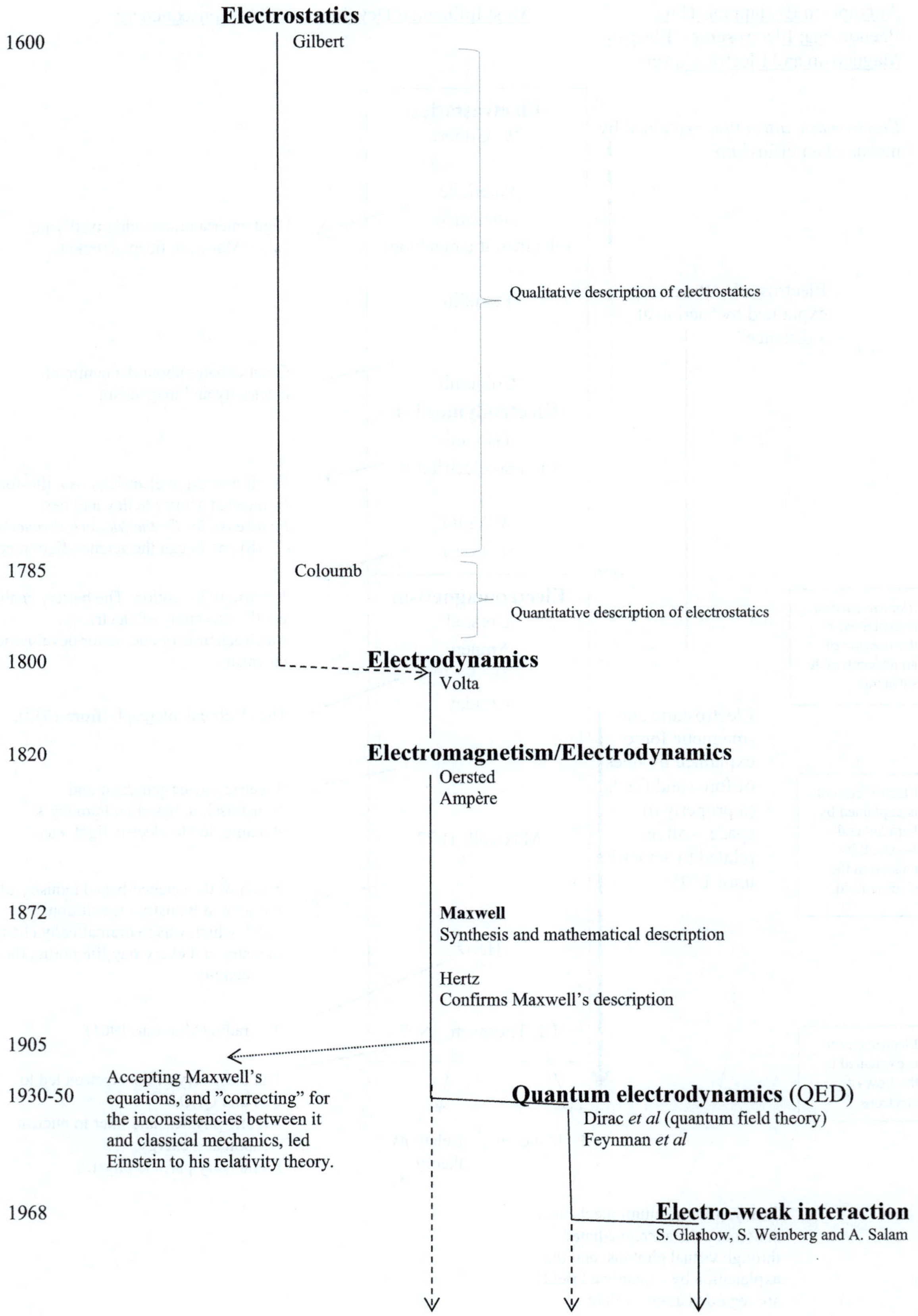

The History of Electricity: 3. Outline

Thales of Miletus, c. 585 BC
Static electricity from amber (*electron* in Greek) was known and discussed. No clear distinction between static electricity and magnetism.

<u>Mechanistic-materialistic explanations</u>

<u>Occult or Aristotelian explanations</u>

Epicureans *et al*
Epicurus, Lucretius and Plutarch attempt materialistic explanation (atomic movements, and/or the flow of air).

The Middle Ages
Electricity and magnetism explained by means of occult (hidden) qualities. Amber can attract because it contains the ability (the hidden quality) to do so.

W. Gilbert, 1600
Distinguished between electricity and magnetism. Attempts materialistic explanation (through effluvium which flows out and in again). Experiments.

Descartes, c. 1635
His mechanistic assumptions of an effluvium to explain magnetism, also influenced views on electricity.

Explanation of attraction in terms of the near occult "action at a distance".

Newton, in his *Opticks* (1704), used both effluvium and "action at a distance" to explain electricity.

Otto von Guericke, c. 1670 and
Francis Hauksbee, c. 1710
Invents and improves first electrostatic generator. Many experiments and public experiments

Many public and private experiments with electricity were made, among others by Gray, Dufay, Nollet, Bose, Kleist and Musschenbroeck, between c. 1650-1750. Little fundamental progress, but many minor developments and led to the discovery of the Leyden jar, c. 1745 (a capacitor), a method of storing and transporting electrostatic charge.

J.A. Nollet, 1745
Produced a one-fluid mechanical theory of electricity. Electricity is the combination of the elements earth and fire – effluent, when streaming away from body, repulsive, and affluent, moving back into the body, attractive.

B. Franklin, c. 1750
Conservation of charge. Lightening is electricity. Explanation of static electricity as surplus (+) and deficiency (-) of amount of unknown (imponderable) fluid, which "acts at a distance".

Ch. Coloumb, c. 1785
Experimental determination of electrostatic force with his new accurate torque scale, and then described the results mathematically.

$$F = k\frac{q_1 q_2}{r^2}$$

(The chart continues on the next page)

Are there more than one kind of electricity?
The beginning of electrodynamics (electricity in motion)

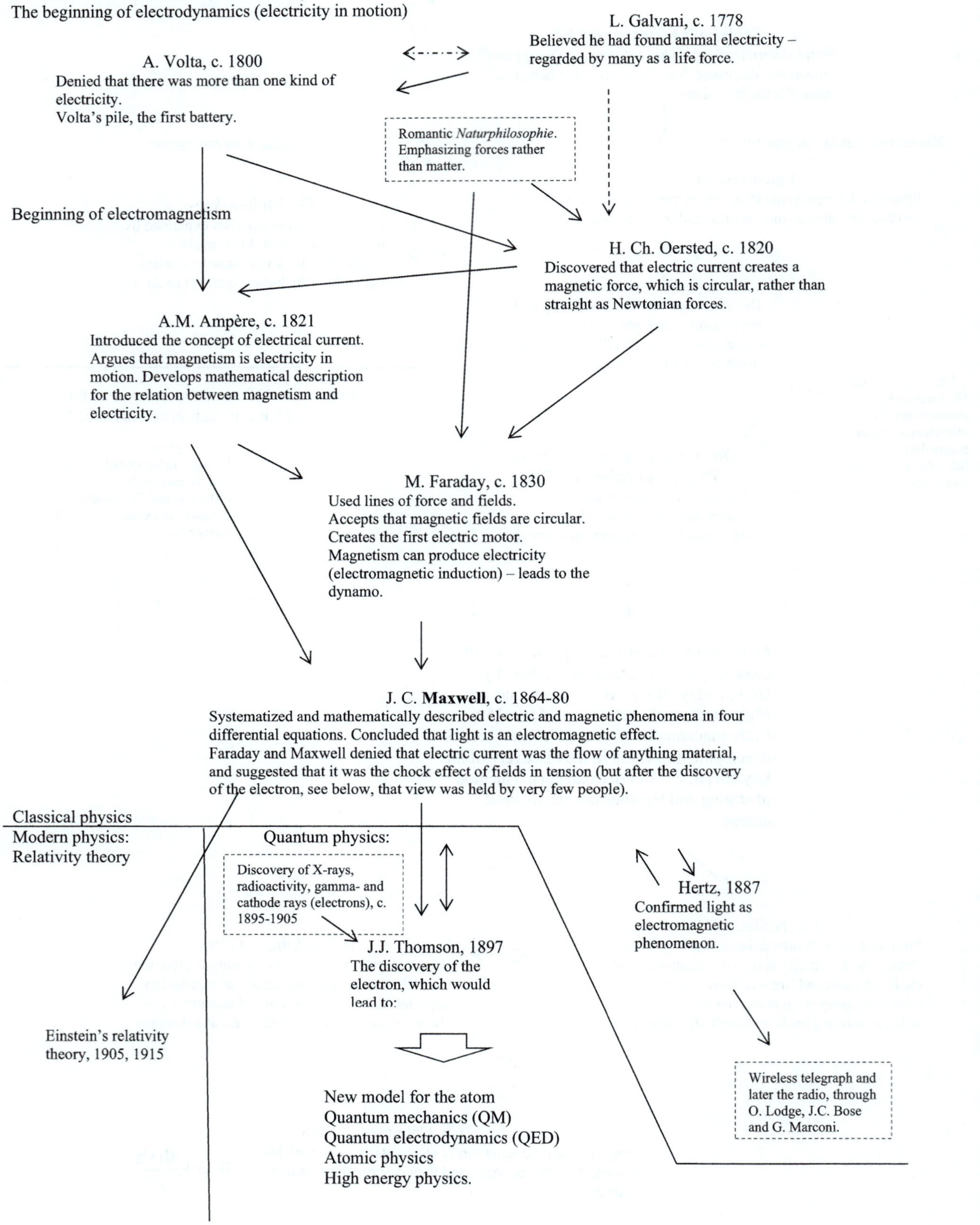

The History of Electricity: 4. More Details

Thales of Miletus, c. 585 BC
Static electricity from amber (*electron* in Greek). No clear distinction between static electricity and magnetism.

(Aristotle)

Epicureans *et al*
Epicurus, Lucretius and Plutarch attempt materialistic explanation (atomic movements, and/or the flow of air).

The Middle Ages
Electricity and magnetism explained by means of occult (hidden) qualities. Amber can attract because it contains the ability (the hidden quality) to do so.

Cardanus, c. 1550
Experiments. Discovers the difference between electricity and magnetism. Attempts materialistic explanation (through vapor and air currents).

W. Gilbert, 1600
Distinguished between electricity and magnetism. Attempts materialistic explanation (through effluvium which flows out and in again. Experiments.

Francis Bacon, c. 1620

Descartes, c. 1635
His mechanistic assumptions of an effluvium to explain magnetism, influenced views on electricity also in terms of effluvium.

Francis Hauksbee, c. 1710
Improved Guericke's machine as a electrostatic generator and made extensive experiments. Shows that it causes blue light and can penetrate glas (but not some fine cloth). Difficult to explain with effluvium.

Otto von Guericke, c. 1670
Invented the first electrostatic machine. Used occult quality (virtue) to explain it – "acting at a distance".

Huygens, c. 1670
Everyone assumed that electrostatic is only attractive, except Huygens, but did not publish or communicate it.

Stephen Gray, 1729
Discovered many new non-solids that can produce static electricity. Discovers the *flow* of static electricity, and distinguished between conductors and isolators (but namned by Jean Desaguliers, c. 1735)

Ch. Dufay and J.A. Nollet, c. 1730
Electrostatic force can be negative, and attempted to find explanation. There must be two kinds of electricity (vitreous and resinous), depending on what was rubbed.

Many public and private experiments with static electricity were made throughout the 18th century, among others by Bose, Kleist and Musschenbroeck, c. 1740, leading to the discovery of the Leyden jar (a capacitor), a method of storing and transporting electrostatic charge.

Leyden jar, 1745

J.A. Nollet, 1745
Produces a theory of electricity and the Leyden jar. Electricity is the combination of the elements earth and fire – effluent, when streaming away from body, repulsive, and affluent, moving back into the body, attractive.

B. Franklin, c. 1750
Conservation of charge.
Lightening is electricity, and suggested ways to study it, incl. the lightening rod. Explanation of static electricity as surplus (+) and deficiency (-) of amount of unknown (imponderable) fluid, which "acts at a distance". Further developed by Aepinus, Priestly and Cavendish.

Ch.A. de Coloumb, c. 1784-89
Experimental determination of electrostatic force with his new accurate torsion balance, and then described the results mathematically. The inverse square law was also independently found by and formulated by Priestly, Cavendish and Robison.

$$F = k\frac{q_1 q_2}{r^2}$$

S. D. Poisson, 1811
Completes the mathematical description of electro- and magneto statics based on a two-fluid theory. Later Gauss formulates it in the modern form.

(The chart continues on the next page)

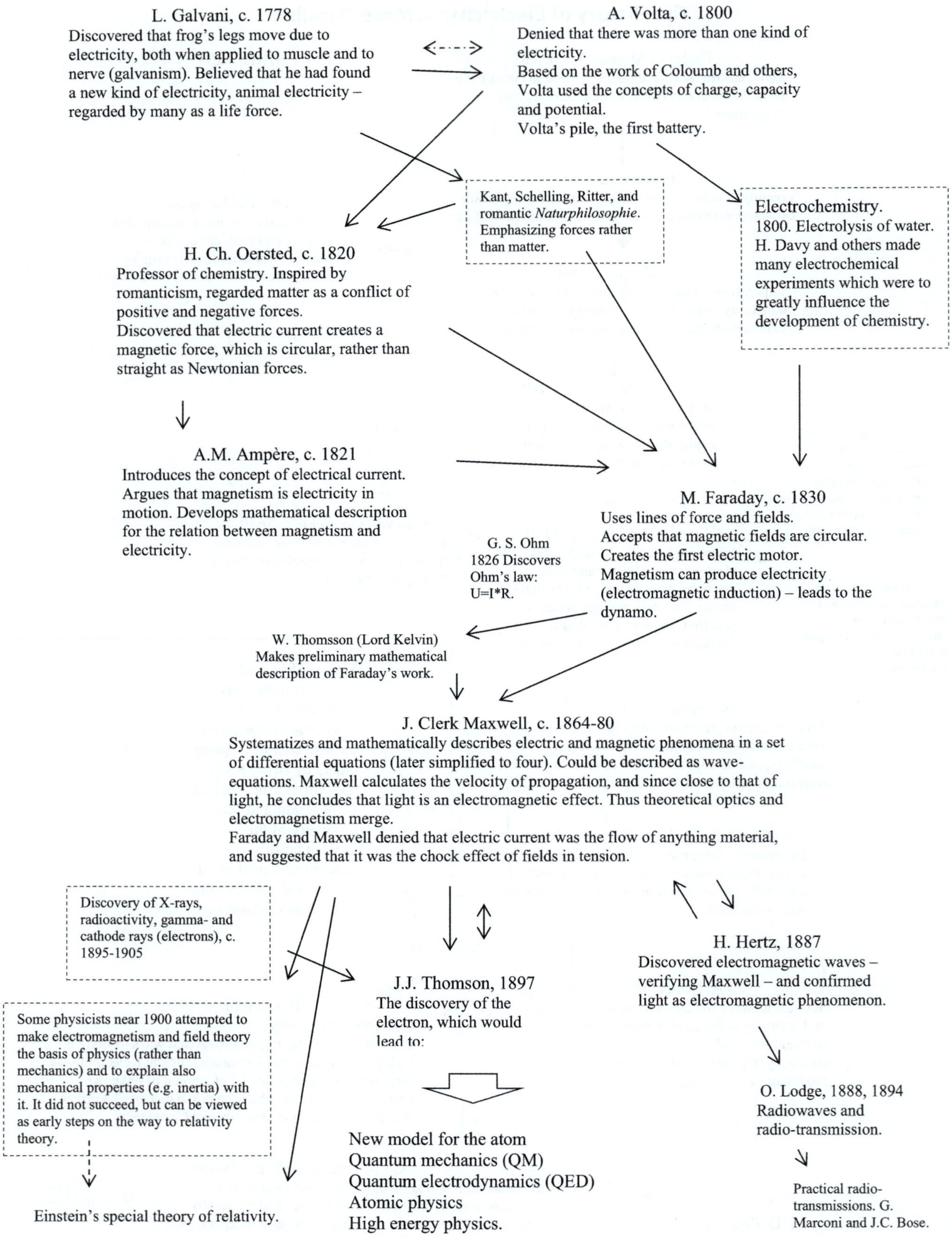
L. Galvani, c. 1778
Discovered that frog's legs move due to electricity, both when applied to muscle and to nerve (galvanism). Believed that he had found a new kind of electricity, animal electricity – regarded by many as a life force.
A. Volta, c. 1800
Denied that there was more than one kind of electricity.
Based on the work of Coloumb and others, Volta used the concepts of charge, capacity and potential.
Volta's pile, the first battery.
Kant, Schelling, Ritter, and romantic *Naturphilosophie*. Emphasizing forces rather than matter.
Electrochemistry.
1800. Electrolysis of water. H. Davy and others made many electrochemical experiments which were to greatly influence the development of chemistry.
H. Ch. Oersted, c. 1820
Professor of chemistry. Inspired by romanticism, regarded matter as a conflict of positive and negative forces.
Discovered that electric current creates a magnetic force, which is circular, rather than straight as Newtonian forces.
A.M. Ampère, c. 1821
Introduces the concept of electrical current. Argues that magnetism is electricity in motion. Develops mathematical description for the relation between magnetism and electricity.
G. S. Ohm
1826 Discovers Ohm's law:
U=I*R.
M. Faraday, c. 1830
Uses lines of force and fields.
Accepts that magnetic fields are circular.
Creates the first electric motor.
Magnetism can produce electricity (electromagnetic induction) – leads to the dynamo.
W. Thomsson (Lord Kelvin)
Makes preliminary mathematical description of Faraday's work.
J. Clerk Maxwell, c. 1864-80
Systematizes and mathematically describes electric and magnetic phenomena in a set of differential equations (later simplified to four). Could be described as wave-equations. Maxwell calculates the velocity of propagation, and since close to that of light, he concludes that light is an electromagnetic effect. Thus theoretical optics and electromagnetism merge.
Faraday and Maxwell denied that electric current was the flow of anything material, and suggested that it was the chock effect of fields in tension.
Discovery of X-rays, radioactivity, gamma- and cathode rays (electrons), c. 1895-1905
J.J. Thomson, 1897
The discovery of the electron, which would lead to:
H. Hertz, 1887
Discovered electromagnetic waves – verifying Maxwell – and confirmed light as electromagnetic phenomenon.
Some physicists near 1900 attempted to make electromagnetism and field theory the basis of physics (rather than mechanics) and to explain also mechanical properties (e.g. inertia) with it. It did not succeed, but can be viewed as early steps on the way to relativity theory.
O. Lodge, 1888, 1894
Radiowaves and radio-transmission.
New model for the atom
Quantum mechanics (QM)
Quantum electrodynamics (QED)
Atomic physics
High energy physics.
Einstein's special theory of relativity.
Practical radio-transmissions. G. Marconi and J.C. Bose.

History of Thermodynamics and Statistical Mechanics 1800–1950: 1. Overview

Thermodynamics, the theory and understanding of heat, was during the first half of the nineteenth century closely linked to the technology of steam engines. Eventually one started to treat heat, especially in relation to gases, statistically, and was able to understand them better. Still later, near 1900, the more abstract statistical mechanics, which analyzed the statistical properties of any ensemble, any collection of mechanical systems was developed.

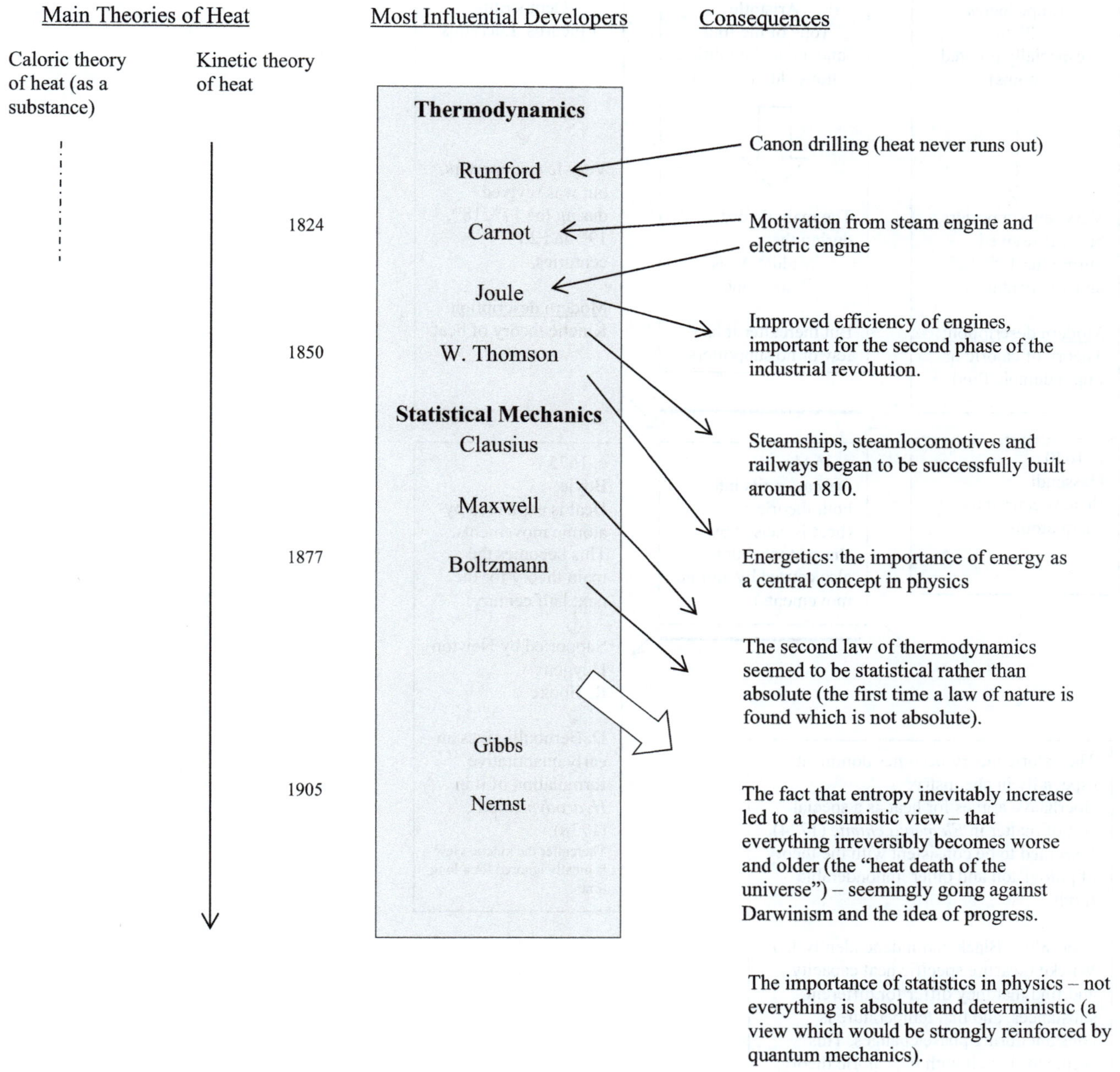

History of Theories of Heat, Thermodynamics and Statistical Mechanics: 2

There were three main theories about fire (heat) in antiquity, which all were historically influential:

1. Heat (fire) as an element

Empedocles
Plato
(especially pointed atoms)

Very few supporters, but was revived during the 17th, 18th and 19th centuries.

Modern description:
Theory of caloric or imponderable fluid

2. Heat (fire) as a primary quality

Aristotle
(one of the four fundamental qualities; hot-cold, dry-wet)

The main theory during the:
Middle Ages
Renaissance

But thereafter it had few or no supporters.

3. Heat (fire) as atomic movements

Democritus, Epicurus, Lucretius

Very few supporters, but was revived during the 17th, 18th, 19th and 20th centuries.

Modern description:
Kinetic theory of heat

c. 1640
Gassendi
(heat is caused by sharp atoms)

c. 1644
Descartes fits into both theories:
(heat is caused by sharp atoms but is also caused by atomic movements).

c. 1675
Boyle
Heat is explained by atomic movements. This becomes the main theory for the next half century:

Supported by Newton
Huygens
R. Hooke

D. Bernoulli gives an early quantitative formulation of it in *Hydrodynamica* (1738).
Thereafter the kinetic view is mostly ignored for a long time

The caloric theory becomes dominant especially in chemistry:
Boerhaave argues for heat as a special sort of matter in *Elementa chimie* (1724). It seemed to be consistent with the theory of phlogiston and other imponderable fluids.

1760-80: J. Black and independently J.G. Wilcke measure specific heat capacity and find that they differ for different substances, and that temperature is constant during phase changes. This seems to fit well with the caloric theory.

Lavoisier and Laplace:
Lavoisier lists caloric as one of the 33 elements in 1789. Laplace constructs complex theory of matter and energy, including caloric.

S.D. Poisson constructs a mathematical formulation of the caloric theory.

(The chart continues on the next page)

c. 1800
Th. Young, H. Davy and Rumford reject caloric in favour of the kinetic theory.

1811. J. Fourier gives a mathematical formulation of (irreversible) heat conduction (but was not accepted and known for a long time).

Sadi Carnot works theoretically on understanding the efficiency of a hypothetical ideal steam engine, *Reflections on the Motive Power of Heat* (1824). He concludes that the motive force produced was due to:
1. the amount of heat
2. the size of the temperature-fall.
Heat not used up, only dissipated.

The end of the caloric theory

James P. Joule, 1843 makes experiments to study the efficiency of actual electric motor. Measured the mechanical equivalent of heat (heat apparently used up and destroyed).
The results consistent with the kinetic theory of heat, but was inconsistent with Carnot's results (primarily because his was reversible while Joule's were irreversible).

The concept "energy" began to be widely used from c. 1855, and a reconstruction of the major fields of physics to allow for energy as a central part of them by e.g. Thomson, Rankine, Helmholtz and Maxwell.

W. Thomson (Lord Kelvin), c. 1850, concluded:
1. Conservation of energy
2. Conservation of heat/temperature (in reversible processes)
3. Defined absolute scale of temperature (Kelvin)

The first law of thermodynamics was also stated or suggested by many others near this time, e.g. Mayer, Helmholtz, Joule and Carnot.

R. Clausius, 1859, concludes:
1. Conservation of energy
2. Entropy (S) strives towards a maximum

He calculated the amount of energy of a gas by adding the kinetic energy of all the molecules – the beginning of statistical mechanics (continued by Maxwell and Boltzmann).

The analyses of Thomson and Clausius led to the rejection of the caloric theory of heat, and a new intensive study of the kinetic theory, which required a statistical theory and analysis (the first in classical physics).

Maxwell, using probability theory, calculates the statistical distribution of velocities of gas molecules in equilibrium, 1859, 66. In 1876 he suggests the thought experiment "Maxwell's demon" to show that it theoretically would be possible that heat could flow from colder to hotter and entropy decrease – which "proved" that entropy and the second law of thermodynamics are not absolute but statistical.

L. Boltzmann continues this work 1877. Succeeds in defining entropy in terms of probability:
$S = k \ln W$ (W = probability states)

Boltzmann's theory failed to give the correct results for calculating specific heats.
However, Einstein, in 1905, using Boltzmann's method, can explain Brownian motion and calculate specific heats.
J.W. Gibbs further generalizes the statistical approach so that it was applicable to any assembly (not just molecules of a gas) in his *Elementary Principles in Statistical Mechanics* (1902).

In 1905 Nernst formulated the third law of thermodynamics, that the entropy tends to zero as the absolute temperature tends to zero (although quantum mechanics later required this to be reformulated). Later a method for determining the absolute value of entropy was developed.

M. Planck uses this approach when he attempts to account for black body radiation 1900

Quantum mechanics

The "Break" Between Classical (Newtonian) and Modern Physics: 1. Overview

During the early 20th century it was discovered that Newtonian physics does not accurately describe some new areas of physics, involving very high speeds and the very small. Two new constants of nature, the speed of light, c, and Planck's constant, h, were discovered, and these are related to and determine much of the physics in these new fields.

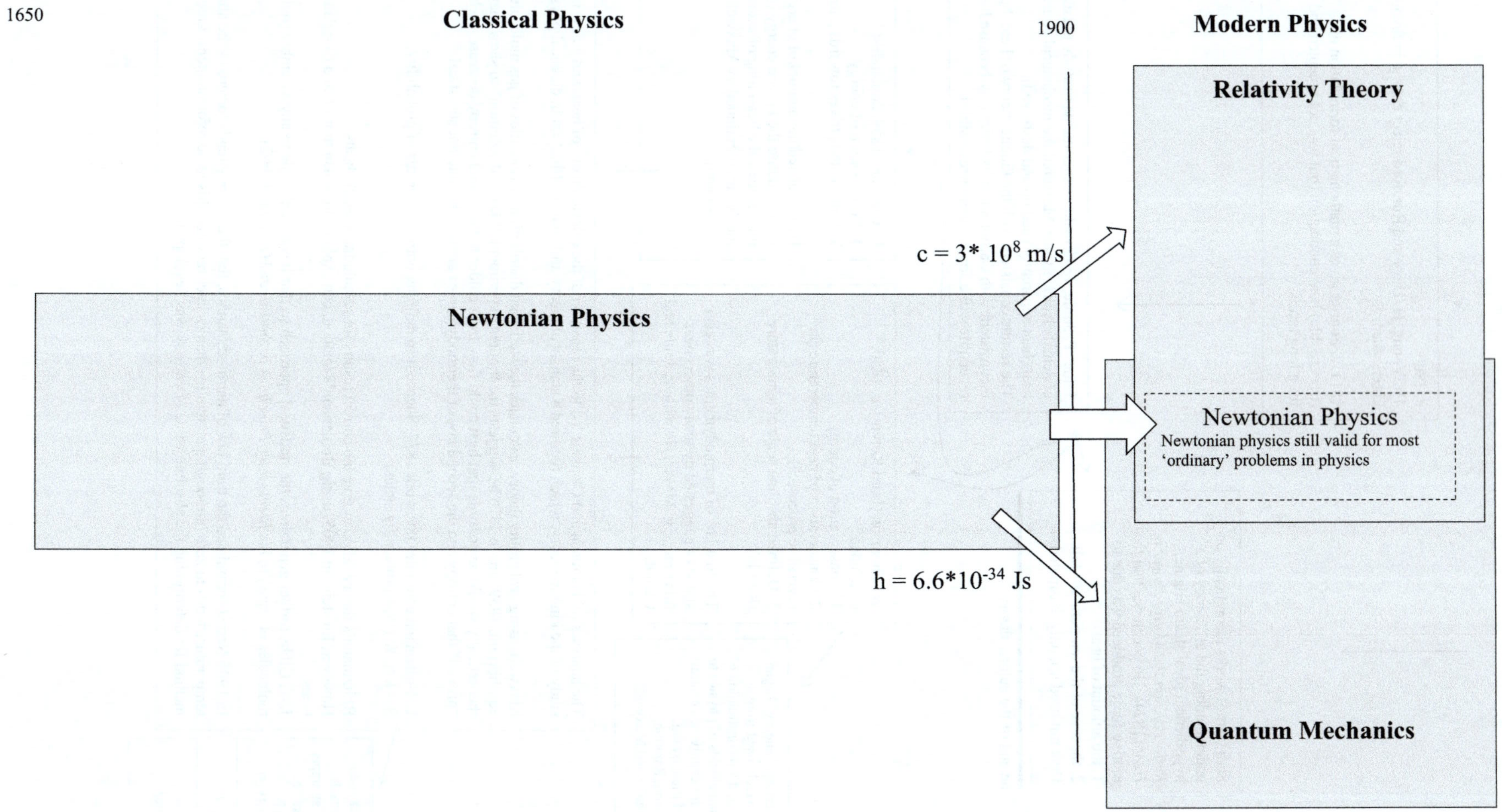

The "Break" Between Classical (Newtonian) and Modern Physics: 2. Physics

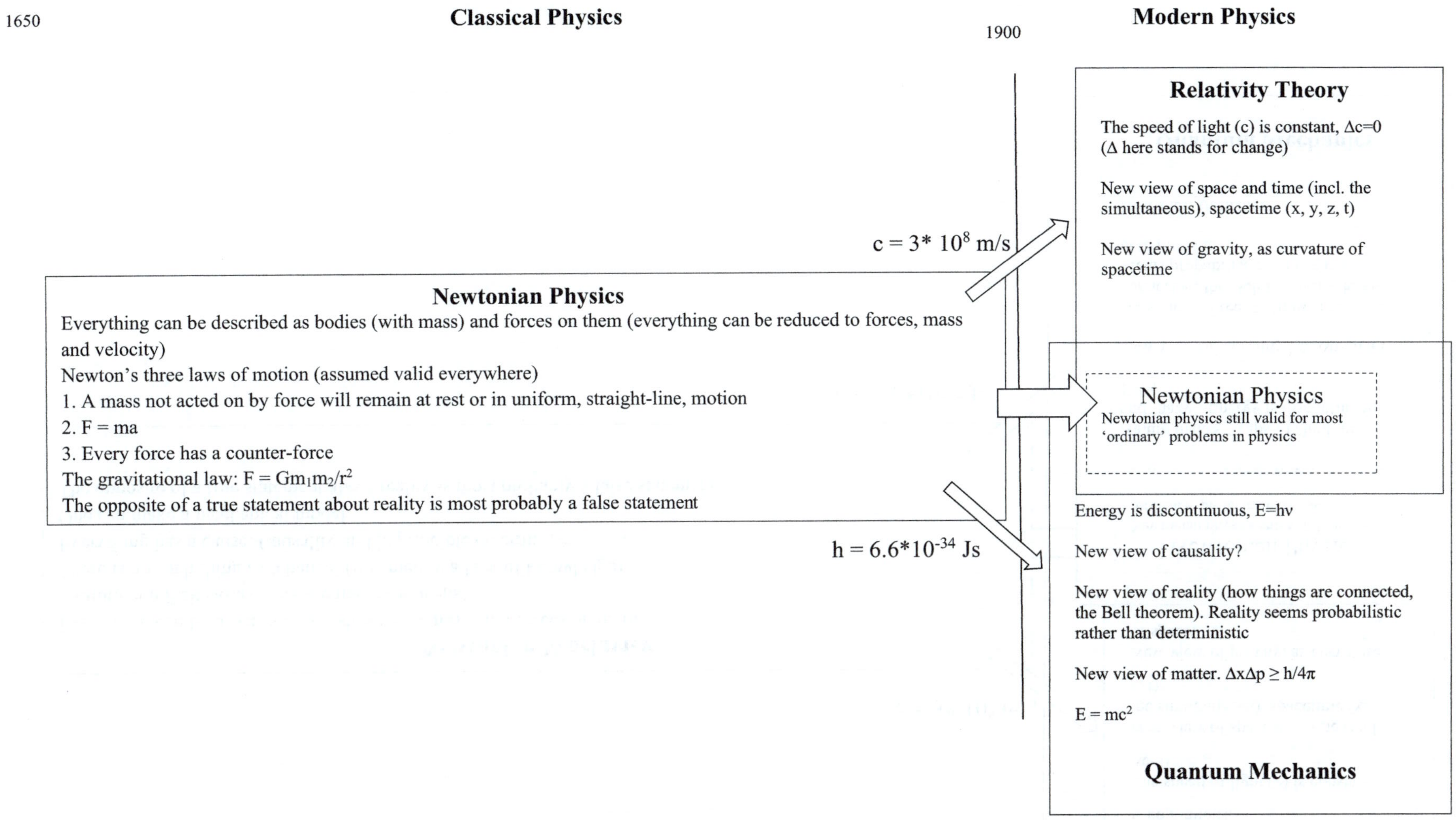

The "Break" Between Classical (Newtonian) and Modern Physics: 3. Conceptionally

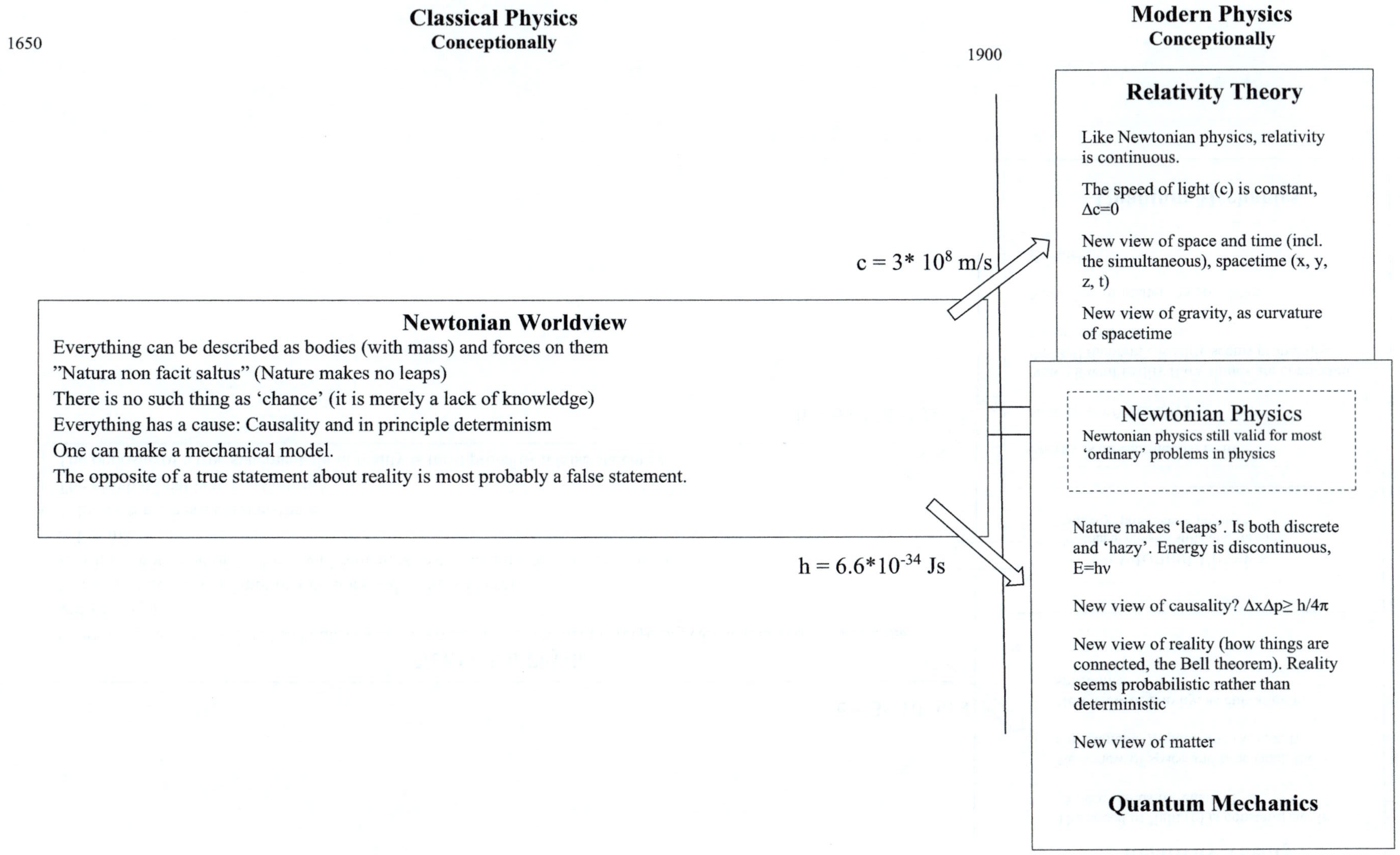

Overview of Modern Physics 1800 – 2020: Main Representatives

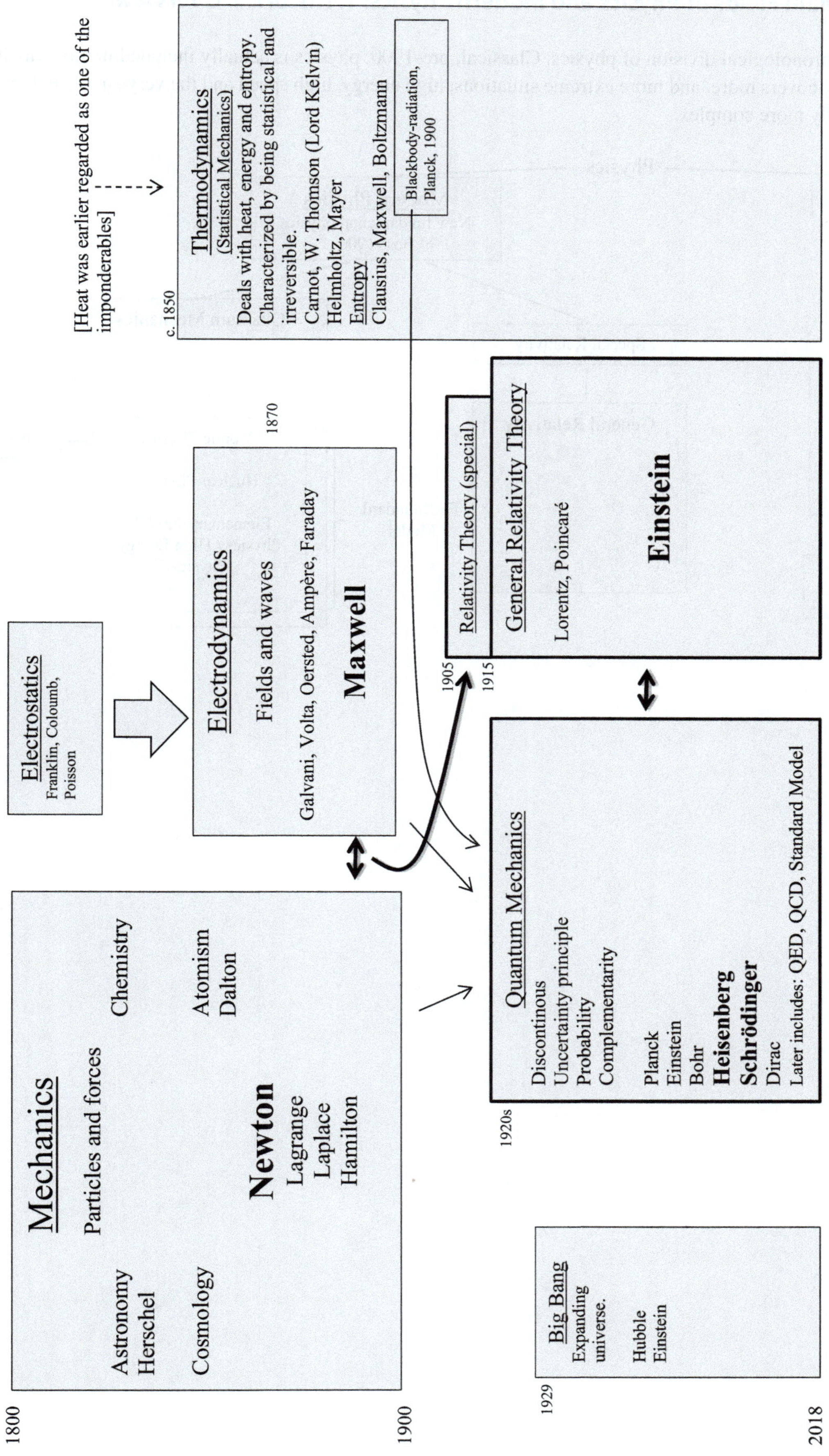

The Main Fields of Physics and Modern Physics: 1. Historical Overview

This chart shows the chronological division of physics. Classical, pre-1900, physics is actually included in modern physics. Modern physics simply covers more, and more extreme situations, high energy, high speed and the very small, and therefore becomes mathematically more complex.

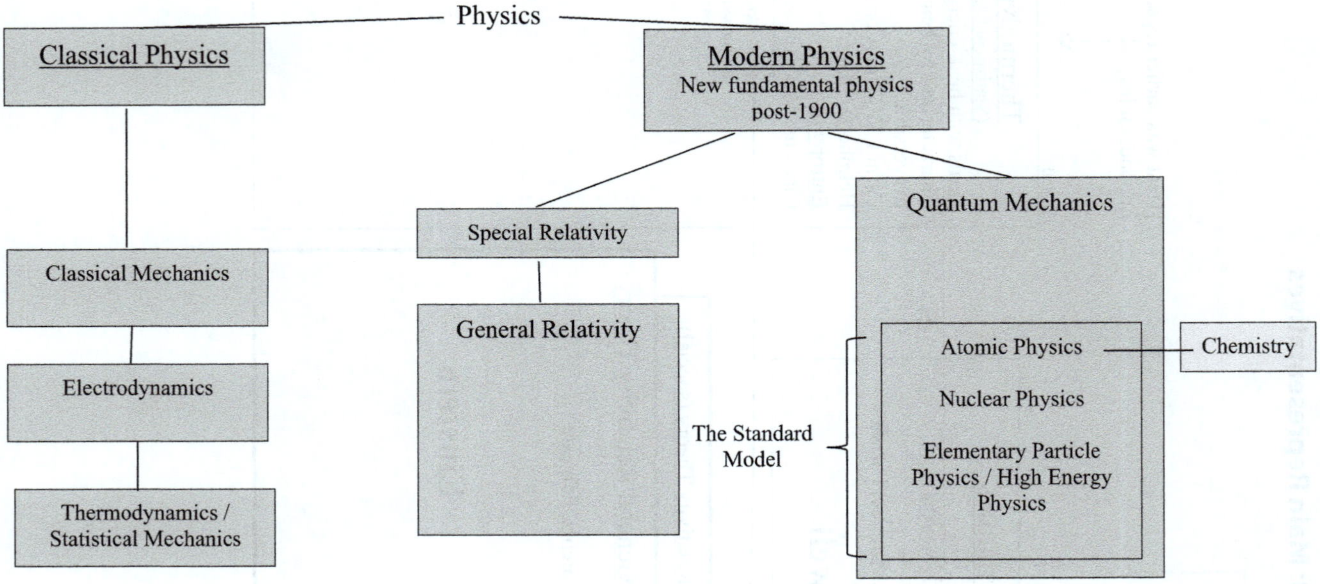

The Main Fields of Physics and Modern Physics: 2. Alternative View

Many physicists regard quantization of energy as that which characterizes modern physics – and therefore they regard relativity theory as a classical (i.e. non-quantized) physics, although it chronologically belongs to modern, i.e. post-1900, physics.

In this chart we also get a better view of the main parts of quantum mechanics, the theories of quantum electrodynamics (of atoms and molecules), of the weak force (involved with radioactivity), which have been unified into electroweak theory, and the strong force in the nucleus (QCD).

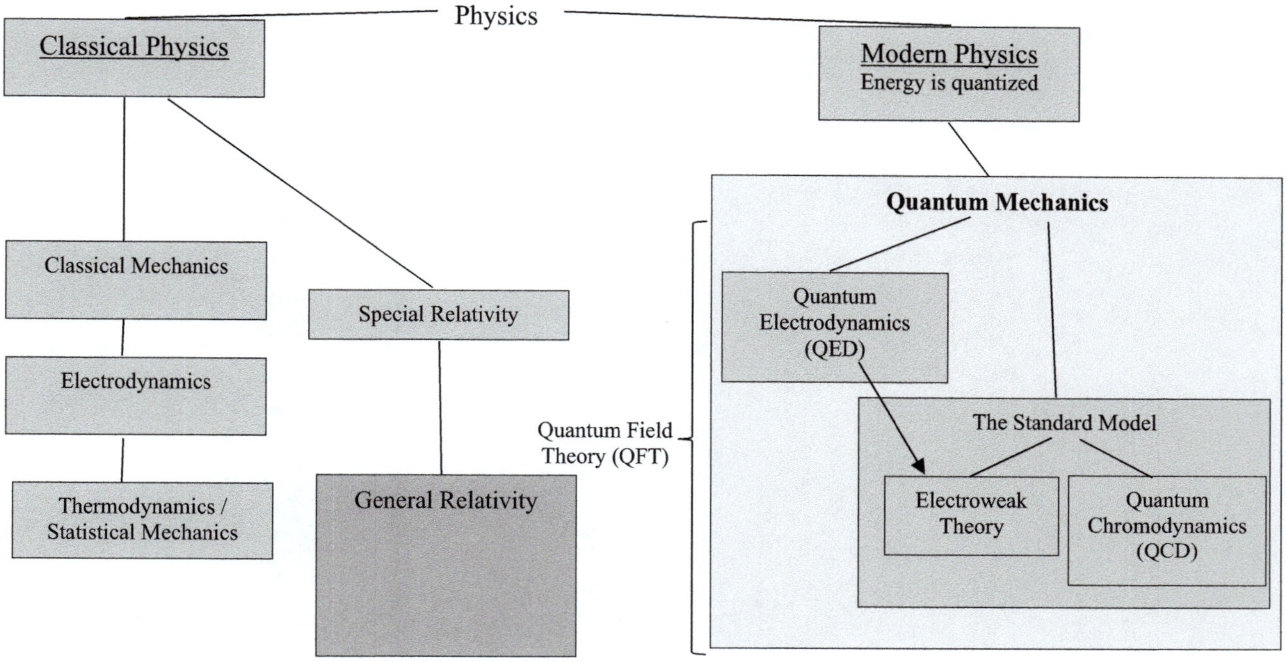

5 History of Chemistry

T. H. Brobjer, *Atlas of the History of Modern Science 1500-2020*, https://doi.org/10.1007/978-3-031-76627-5_5

Overview of the History of Chemistry (1600–2000)

Close to Chronologically Proportional

The distinction between organic (carbon-based, associated with living organisms) and inorganic chemistry only began to be used in the late eighteenth century, and did not exist in the early parts of this time-period.

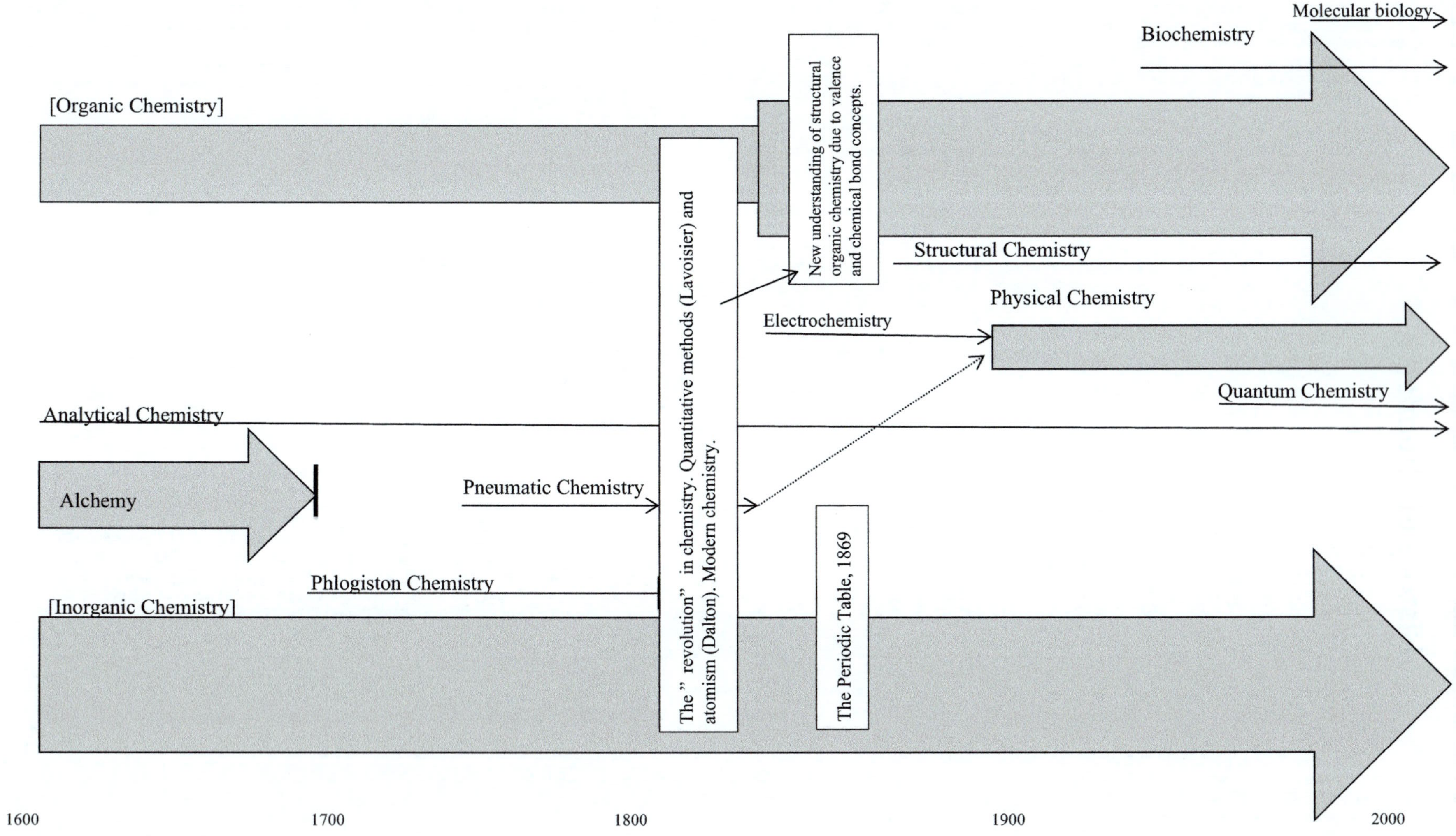

Map of the History of Chemistry 1500–1950: Overview

When chemistry – a late science compared to physics, mathematics and biology – became more established, its origin and sources were manifold:
The word "chymistry" is often used to signify the subject during the time when it is impossible to entangle it from alchemy.

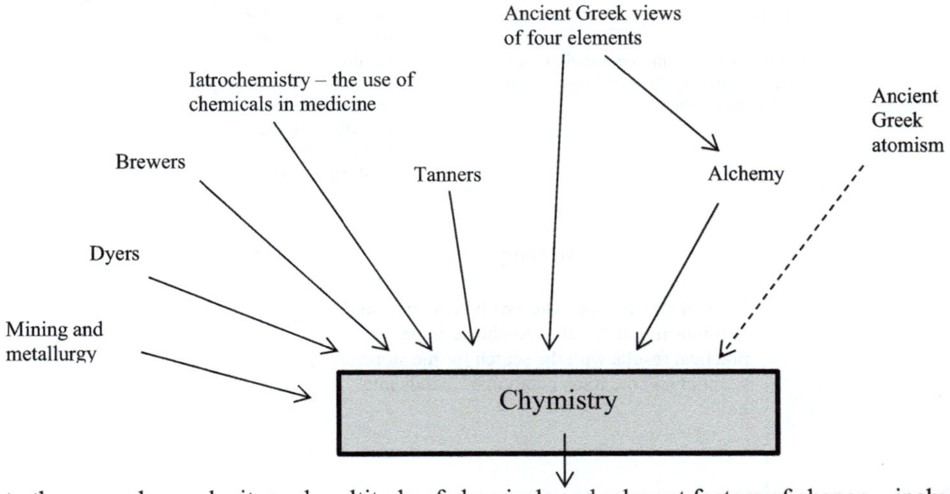

Due to the general complexity and multitude of chemicals and relevant factors of change – including the obscurity and absence of a common language, absence of criteria of purity, uncertainty as to what the fundamental units were (the elements) and the lack of knowledge of gases (and thus that which went away as gas simply was assumed to have disappeared) chemistry developed very slowly until c. 1750 when pneumatic chemistry began to make progress (based on early work on the air pump, with Boyle and others).

↓

Better understanding of gases and the **discovery of oxygen**, 1773/74 (necessary for a correct understanding of combustion and oxidation)

↓

Lavoisier creates modern quantitative chemistry, c. 1780-90

↓

Dalton and others propose **the chemical atomic theory**, c. 1810

↓

Dramatic expanse of chemistry, and especially organic and structural chemistry during the 19th century

↓ ↘ Biochemistry

Mendelejev's periodic table, c. 1869, enables new ways of understanding and organizing chemistry (especially inorganic chemistry)

↓

Quantum chemistry, c. 1930-40, makes it possible for the first time to understand chemical bonding (three sorts of bonds)

[Chemistry only became a modern science with Lavoisier at the end of the eighteenth century, so the rest of chemistry is treated below under the development of science during the nineteenth and twentieth centuries.]

Map of the History of Chemistry, c. 1000 B.C. – A.D. 2000

Chronologically Proportional

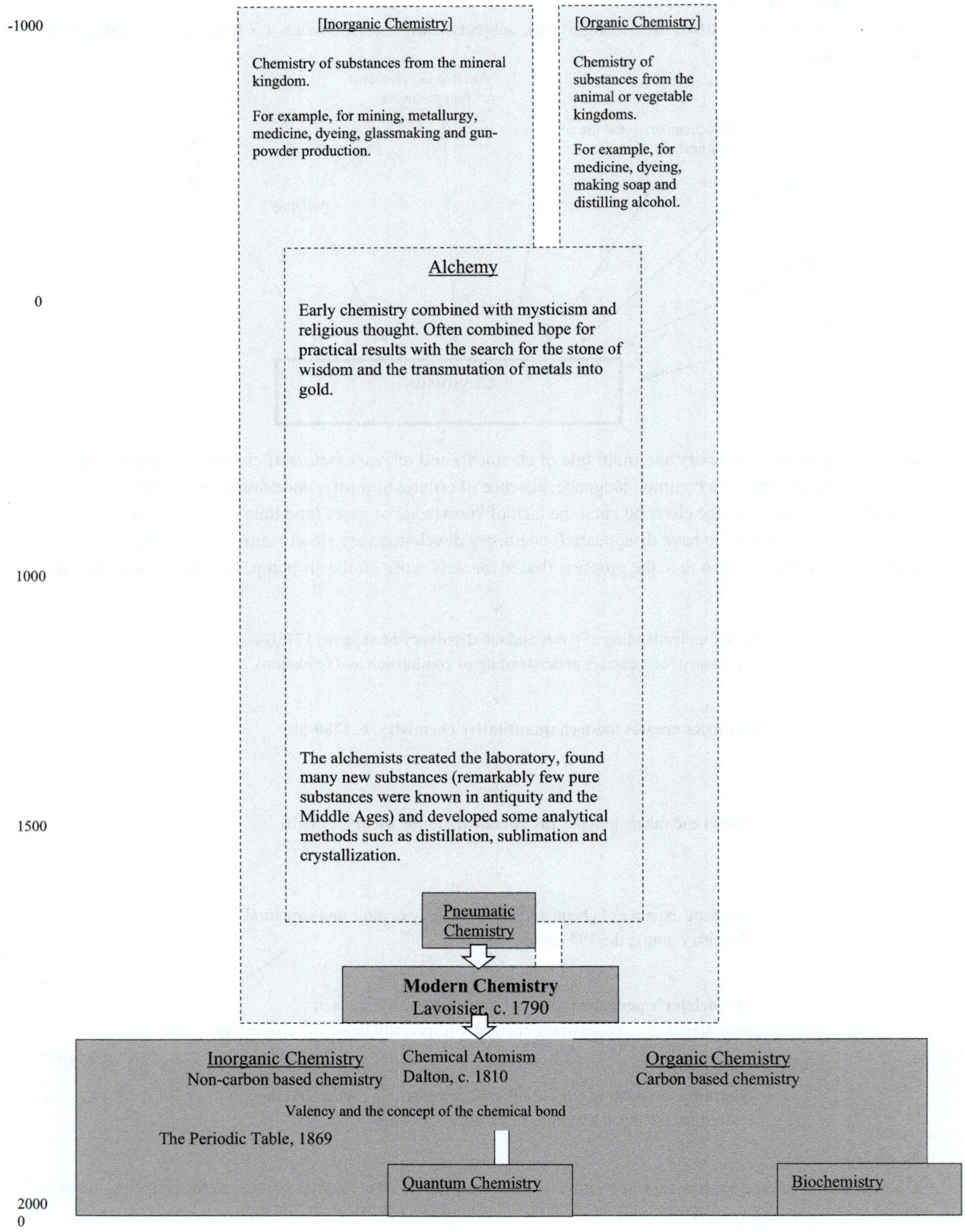

History of Chemistry 1500–1970: 1. Chemists

Topological Map of Chemists

Theoretical, inorganic and physical chemistry | Organic and biochemistry

Paracelsus, 1530

van Helmont, 1620

Gassendi, 1630

Boyle, 1670 Becher, 1667

Stahl, 1697

Newton, 1704

Geoffroy the Elder, 1710

Black, Cavendish, 1770
Scheele, Priestley, 1774

Lavoisier, 1780s

Proust, 1800 <-> Berthollet, 1800

Dalton, 1808 Gay-Lussac, 1808
Davy, 1810 **Berzelius**, 1810
Avogadro, 1811

Faraday, 1830 Wöhler, 1828
Dumas, 1840 Liebig, 1840
Pasteur, 1848

Bunsen, Kirchhoff, 1859
Cannizzaro, 1860 **Frankland**, 1852, 66 **Kekulé**, 1858
Clausius, 1865

Mendelejev, 1869

Maxwell van't Hoff, 1874
Le Bel, 1874

Boltzmann, 1877 Gibbs, 1877 Arrhenius, 1877
A. Werner, 1893 Ostwald, 1890

J.J. Thomson, 1897
Rutherford, 1911 Marie Curie, 1910 Nernst, 1905 E. Fischer, 1900
Bohr, 1913

Lewis, 1915
Langmuir, 1920 Hund, Hückel, Mullikan, 1920s
Pauling, 1925

Miller-Urey, 1953
Pauling, 1950
Watson-Crick, 1953

History of Chemistry 1500–1970: 2. With Some Relations

Topological Map of Chemists

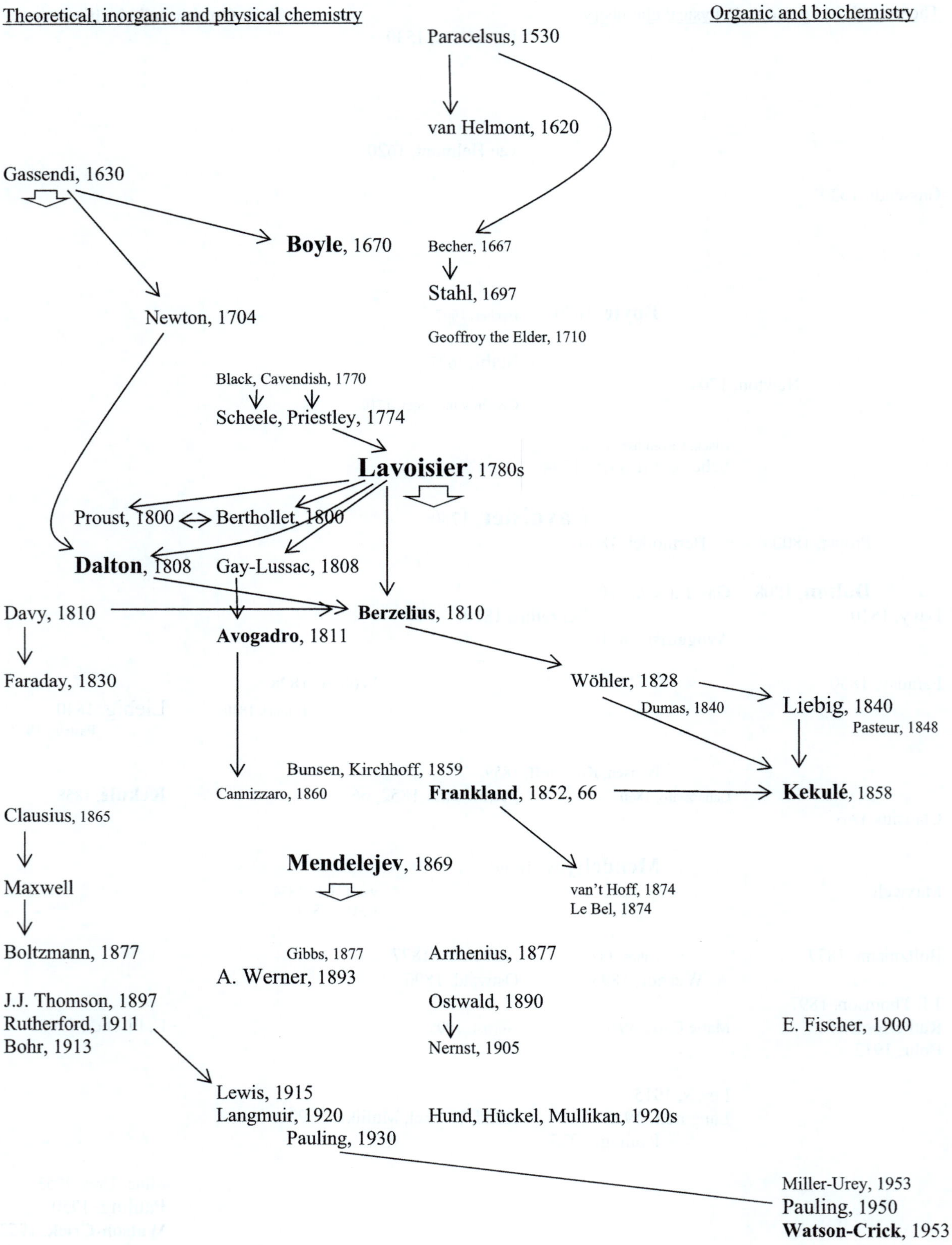

History of Chemistry 1500–2000: 3. With Brief Summaries

Topological Map of Chemists

Tending toward:
Theoretical, inorganic and physical chemistry
1500

Tending toward:
Organic and biochemistry

Paracelsus, 1530
Important alchemical thinker and practician. Founded iatrochemistry (the use of chemistry to medicine). Three principles (*Tria prima*).

van Helmont, 1620
Important alchemical thinker and practician. Worked on air. Air and fermentation

Gassendi, 1630
Important for the revival of ancient (Epicurean) atomism

Boyle, 1670
Study of air. Sceptical and empirical.
Rejected the four-element and *tria prima* theory of matter.
Corpuscular theory.

Stahl, 1697
Influential broad chemist. Vitalism. Phlogiston

Newton, 1704
Atomism. Alchemy. Imponderable fluids
Influenced chemical explanations in terms of elementary corpusculars, masses, affinities and imponderable fluids for at least two centuries

Geoffroy the Elder, 1710
Published tables of affinities

Black, Cavendish, 1770
Discovered new gases ("airs"), H_2 and CO_2

Scheele, Priestley, 1774
Discovered oxygen ("dephlogisticated air")

Lavoisier, 1780s
Quantitative methods. Oxygen instead of phlogiston. New nomenclature.

Proust, 1800 ⟷ Berthollet, 1800
The law of definite proportions. | Indefinite (variable) proportions.

Dalton, 1808
Atomism. Elements. First table of atomic weights. The law of multiple proportions.

Gay-Lussac, 1808
Reacting gas volumes are always in simple proportions.

Davy, 1810
First to use electrolysis to identify six new elements

Berzelius, 1810
The most influential chemist in the first half of the century.
Developed atomism, and a dualistic electrochemical theory, and determined accurate relative atomic weights

Avogadro, 1811
The Avogadro hypothesis (equal volumes contains equal number of molecules).
The concept of "mole".
This could be used to determine relative molecular weights.
The molecules of common elementary gases are composite.

Faraday, 1830
Electrochemistry

Wöhler, 1828
Influential chemist. Synthesis of urea.

Liebig, 1840
Highly influential organic chemist and teacher.
Developed analysis of organic chemistry.

(The chart continues on the next page)

Pasteur, 1848
Discovered chirality

Clausius, 1857
Kinetic theory of gases.

Bunsen, Kirchhoff, 1859
Discovered and developed spectrocopy

Cannizzaro, 1860
Revival of Avogadro's work.
Clarified the distinction between atoms and molecules.

Frankland, 1852, 66
Developed the concept of valence and chemical bond.

Kekulé, 1858
Highly influential organic chemist. Structural chemistry. Benzene.

Mendelejev, 1869
Discovered the periodic system. New elements.

Maxwell
Further developed the kinetic theory of gases.

van't Hoff, Le Bel, 1874
Independently, suggested carbon's tetrahedral structure.

Boltzmann, 1877
Developed kinetic gas theory. Enthropy.

Gibbs, 1877
Free energy.
Phase rules.

Arrhenius, 1877
Ion chemistry

A. Werner, 1893
Coordination chemistry

J.J. Thomson, 1897
Discovery of the electron

Ostwald, 1890
Many aspects of physical chemistry. Sceptical of atomism.

Rutherford, 1911
Planetary atomic model.

Nernst, 1905
The third law of thermodynamics

E. Fischer, 1900
Many aspects of organic and biochemistry

Bohr, 1913
The old quantum chemistry

Marie Curie, 1910
New radioactive elements

Lewis, 1915
Suggested electron pair sharing

Langmuir, 1920
Developed early concepts of chemical bonding

Hund, Hückel, Mullikan, 1920s
Developed early concepts of chemical bonding

Miller-Urey, 1953
Experimental production of amino acids from simple molecules.

Pauling, 1950
Many aspects of chemistry and biochemistry
Early development of chemical bond and benzene.
Protein structure, incl. alpha helix.

Watson-Crick, 1953
The structure of DNA

History of the Theories and Views of the Fundamental Units of Matter

The overall views of the fundamental units of matter (ignoring here the view of matter as immaterial) can be divided into essentially four kinds and stages.

Atomism
Very large number of different sorts of fundamental units, atoms.

The Four-Element View
Empedocles, Aristotle and many others. The standard view until c. 1670. Everything consists of mixtures of the four elements; earth, fire, water and air.

Alchemy
Different views, but most influential (apart from the four-element theory) was Paracelsus' *tria prima*; everything consists of salt, sulphur and mercury. Three substances or principles, sometimes combined with the four-element view.

Chemical Elements
Lavoisier and Modern Chemistry, 1790
Lavoisier listed 33 elements (but some of them are not regarded as such today).

Many New Elements Discovered:
By the end of the nineteenth century the number of known elements had risen to c. 80 (and "false" elements had been discarded).

Attempts at Simplification:
There existed theories which attempted to simplify this, including that all the elements were combinations of hydrogen (W. Prout, 1815) – but that was discarded when it did not seem to agree with atomic weights.

Elementary Particles
The Discovery of the Electron, 1897:
Atoms were not the fundamental units. The proton was found in 1917, and for a period it seemed that perhaps everything was built from only these two particles.

The Discovery of Other Elementary Particles:
In 1932 the neutron and the positron was found. Soon many other particles were found. By the 1960s hundreds of "elementary" particles had been found, but no system.

The Development of the Standard Model:
In the 1960s and 70s, quarks were suggested as more fundamental particles, and the standard model developed.

Today's View of Elementary Particles:
Today 17 elementary particles have been identified in the completed Standard Model (or 61, depending on how they are counted). The existence of "dark matter" strongly suggests further units.

History of the Theories and Discoveries of the Elements

The views of how many and what are the elements – the substances which cannot be broken down further by chemical reactions. Until one had a reasonable view of the fundamental units of matter and chemistry, it was difficult for the field of chemistry to advance significantly.

Atomisms
Democritus, Epicurus and Lucretius

There are a large number of fundamental units of matter, atoms.

The Four-Element View
Empedocles, Aristotle and many others. The standard view until c. 1670.
Everything consists of mixtures of the four elements; earth, fire, water and air.

Historically Known Elements
The views of the fundamental units of matter (elements) were highly uncertain and nebulous until c. 1790 (Lavoisier).

The main views and theories were the four-element theory, and alternative alchemical theories.

However, with hindsight, we can note that since antiquity nine elements were known, seven metals (same as the number of heavenly bodies); gold, silver, copper, iron, tin, lead and mercury, and carbon and sulphur. Four further elements were known since the Middle Ages; arsenic, antimony, bismuth and zink.

Alchemy
Alchemy was often based on the four-element view (but other views also existed).

Paracelsus, c. 1525
Paracelsus introduced an influential alternative view: the *tria prima*: everything consists of mixtures of salt, sulphur and mercury (sometimes regarded as elements, sometimes as principles).

Helmontianism
J.B. van Helmont argued that there were only two fundamental units; water and air (or ferment), c. 1620.

The Discovery of New Elements after c. 1650
There was a slow discovery of different elements (and what we today regard as false elements, such as phlogiston, heat, light, etc), including phosphorus (c. 1670), nitrogen (1772), hydrogen (1766-81), etc.

Because of the importance of oxidation processes (e.g. combustion and respiration), it was the discovery of oxygen in 1773/74 which was most important.

Lavoisier and Modern Chemistry, c. 1790
Lavoisier gave a definition and list of known elements.

Definition: A substance that cannot be broken down further.

List of elements: 33, including light and heat (but phlogiston had been rejected).

Four categories: Gases, non-metals, metals and earths.

By 1830, 55 elements were known (of the c. 92 naturally occurring elements). One sought and found some periodicity among them.

Important for the discovery of many new elements during the beginning of the nineteenth century were electrolysis, and by the middle of the century, spectroscopy.

The Periodic Table, 1869. Mendelejev.
Mendelejev listed 63 elements and predicted further ones – and new unpredicted ones were soon discovered. The table was during this time listed according to atomic weights. Only after c. 1913 did one begin to list it according to atomic numbers.
By c. 1950 close to one hundred elements identified, incl. short-lived radioactive elements and elements made by humans (with atomic number above 92).

History of the Concept of the Molecule and Molecules

Central to the understanding of chemistry and matter is not only the concept of atoms and elements, but also the concept of chemical compounds. The concept of the molecule, units of several atoms bound together, was slow to develop, and remained very vague (which basically also was true for the concept of the atom) until the elements were well known and classified and a theory of valence and chemical bond was established (c. 1860s).

The word "molecule"
The term molecule (smallest mass) comes from P. Gassendi's treatise on atomic theory, *Syntagma Philosophicum* (1660). The term was adopted by atomists (corpuscular philosophers) and entered into all European languages. However, it was vaguely used, and made little progress until atomism was better developed.

↓

Dalton's atomism, 1803, 1810

↓

Stoichiometric considerations.
Regularities in chemical reactions (based on the work of J.L. Proust, Gay-Lussac, *et al*) seemed to imply atomism and molecules (with definite number of combined atoms).

Berzelius and Electrostatic force.
Berzelius developed the atomic theory, and presented a polarized electrical explanation for chemical binding (but which could not conceive of molecules between identical atoms, e.g. H_2).

↓

Theories of molecules
By the 1840s there were several views and theories of how molecules formed, which competed.

↓

The Laurent-Gerhardt View
The Laurent-Gerhardt theory – which basically is the same as that held today – won in the 1850s. It could explain isomeric compounds.

The Avogadro Hypothesis, 1811
The same volume of gas contains the same number of particles. To explain experimental results (e.g. that the compound of oxygen and hydrogen, water, weighed less than the oxygen alone [O_2]) Avogadro assumed that some elements divide into two when reacting ($O_2 \rightarrow 2O$).
This seemed improbable and could not be reconciled with views of chemical binding (as electrical) – and the hypothesis was ignored for half a century.
The same hypothesis was also proposed by Ampère, shortly after Avogadro.

↓

Cannizzaro's revival of Avogadro, 1860

The Acceptance of the Concept of Molecule
By the 1860s there was a general consensus of the existence of and general structure of molecules, as well as about their actual sizes.
[But there was no understanding of the nature of bonding – see the discussion of quantum chemistry below.]
This development went hand in hand with the great success of structural chemistry at this time.

Chemistry and Worldviews, c. 1700–2000

Chemistry has been more closely connected to material production, including industry, than to generating worldviews than has been the case for physics and biology. Chemistry has changed how we live – plastics, medicines, dye-stoff, etc – but less how we conceive the world. However, during the seventeenth century it aided in bringing forth the scientific revolution, and in the nineteenth century, it strongly favoured a materialistic (and atomistic) view of the world.

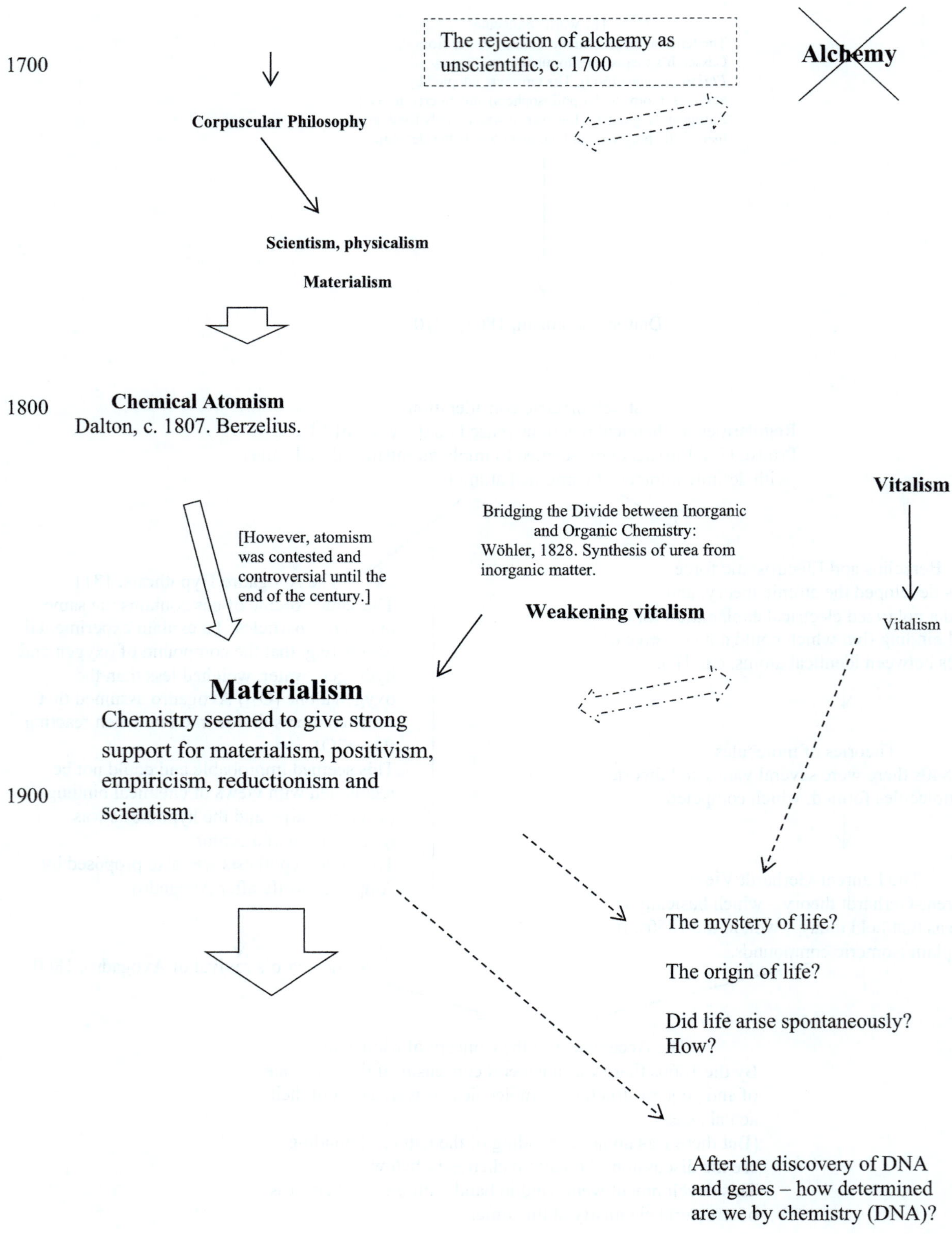

History of Chemistry and Its Major Subfields, c. 1600–2000: 1. Overview

Chronologically Proportional

The different subfields of chemistry overlap more than the subfields of physics and biology, which is not always possible to well illustrate. Analytical chemistry, the methods of performing chemical analysis, i.e. determining substances, is here not given its own field since it is important in all chemical fields (see also the third chart in this series below). The distinction between organic (carbon-based, associated with living organisms) and inorganic chemistry only began to be used in the late eighteenth century, and did not exist in the early parts of this time-period.

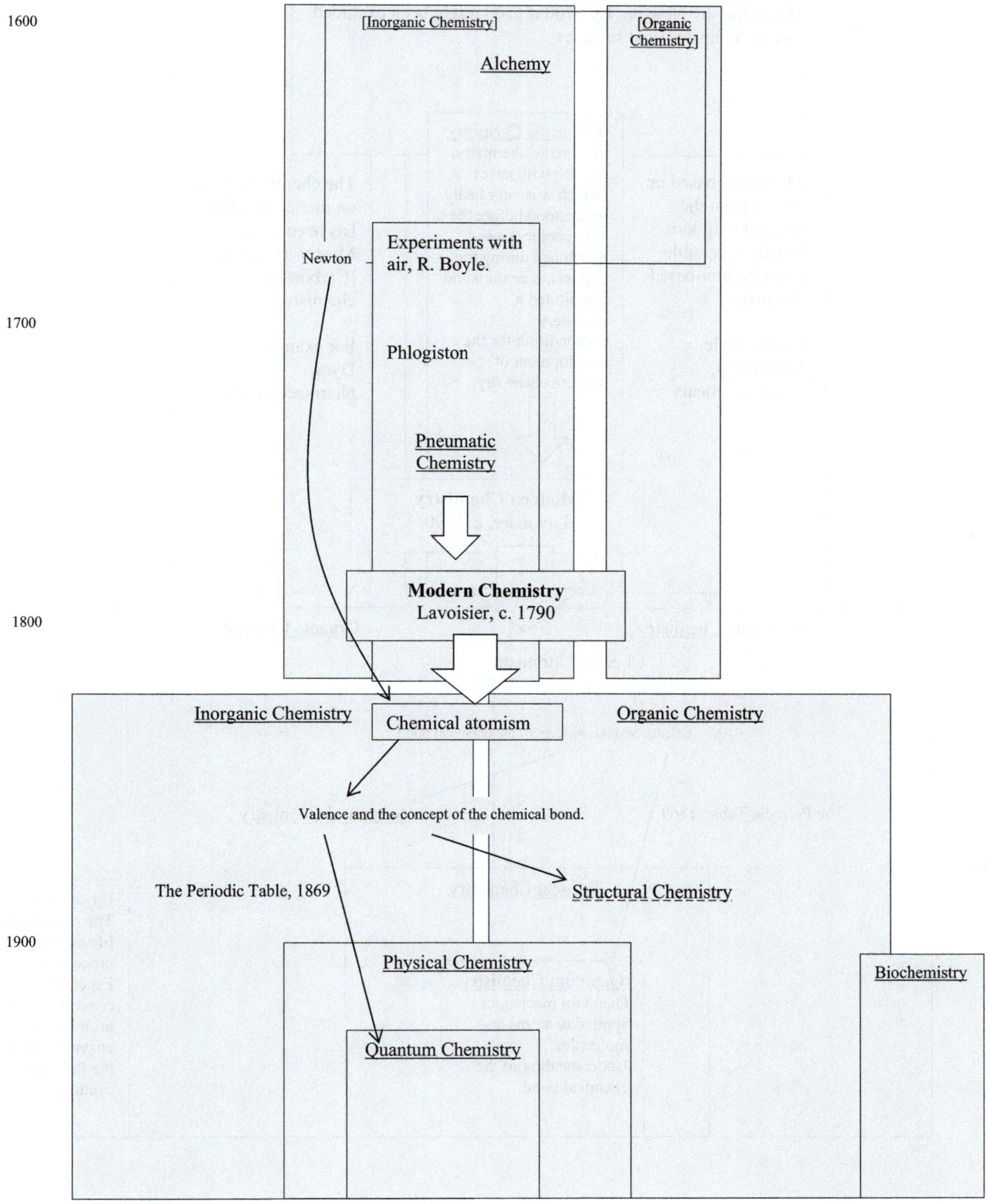

History of Chemistry and Its Major Subfields, c. 1600–2000: 2. Descriptions

Chronologically Proportional

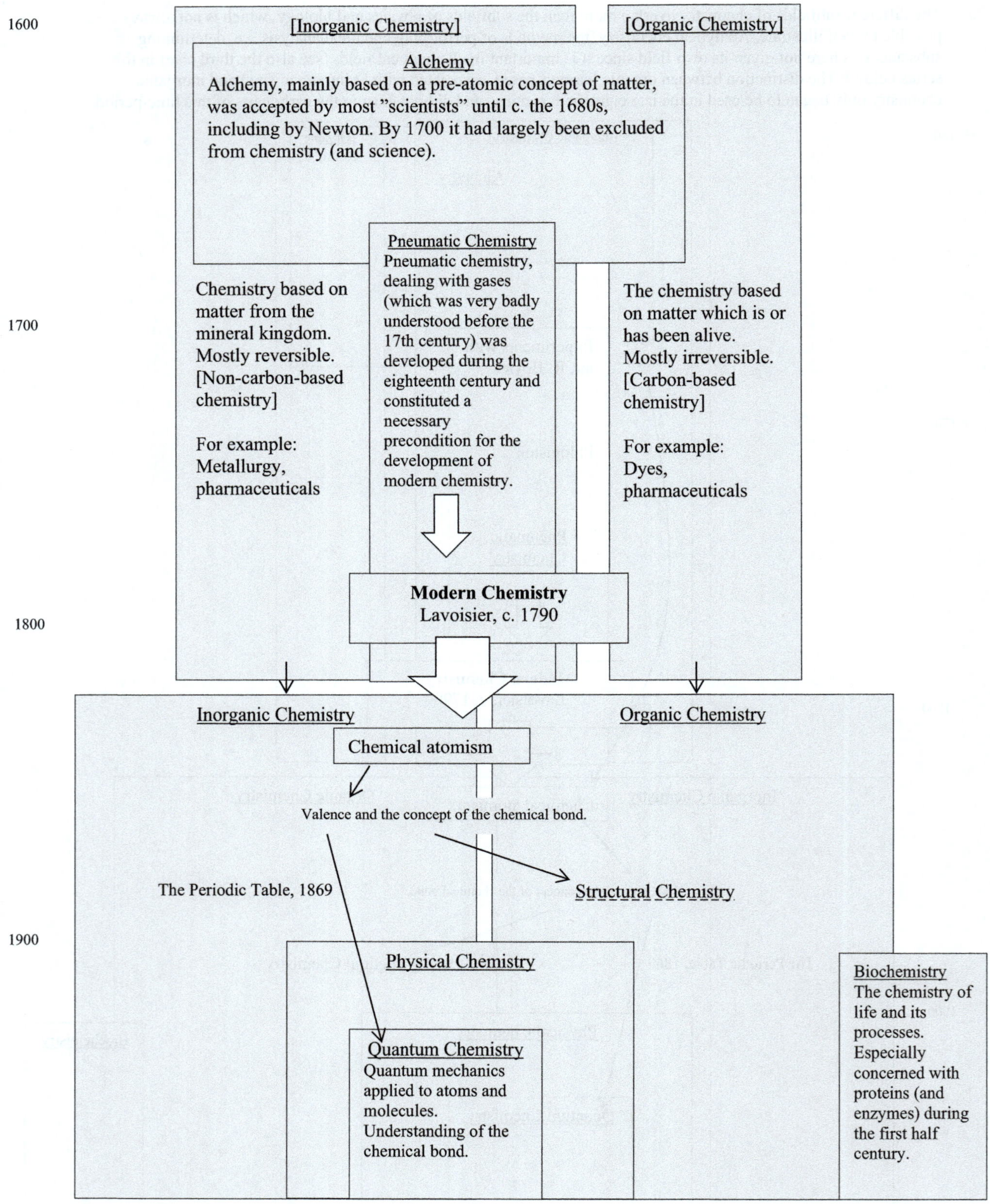

History of Chemistry and Its Major Subfields, c. 1600–2000: 3. Analysis

The Development of Methods and Instruments of Analysis

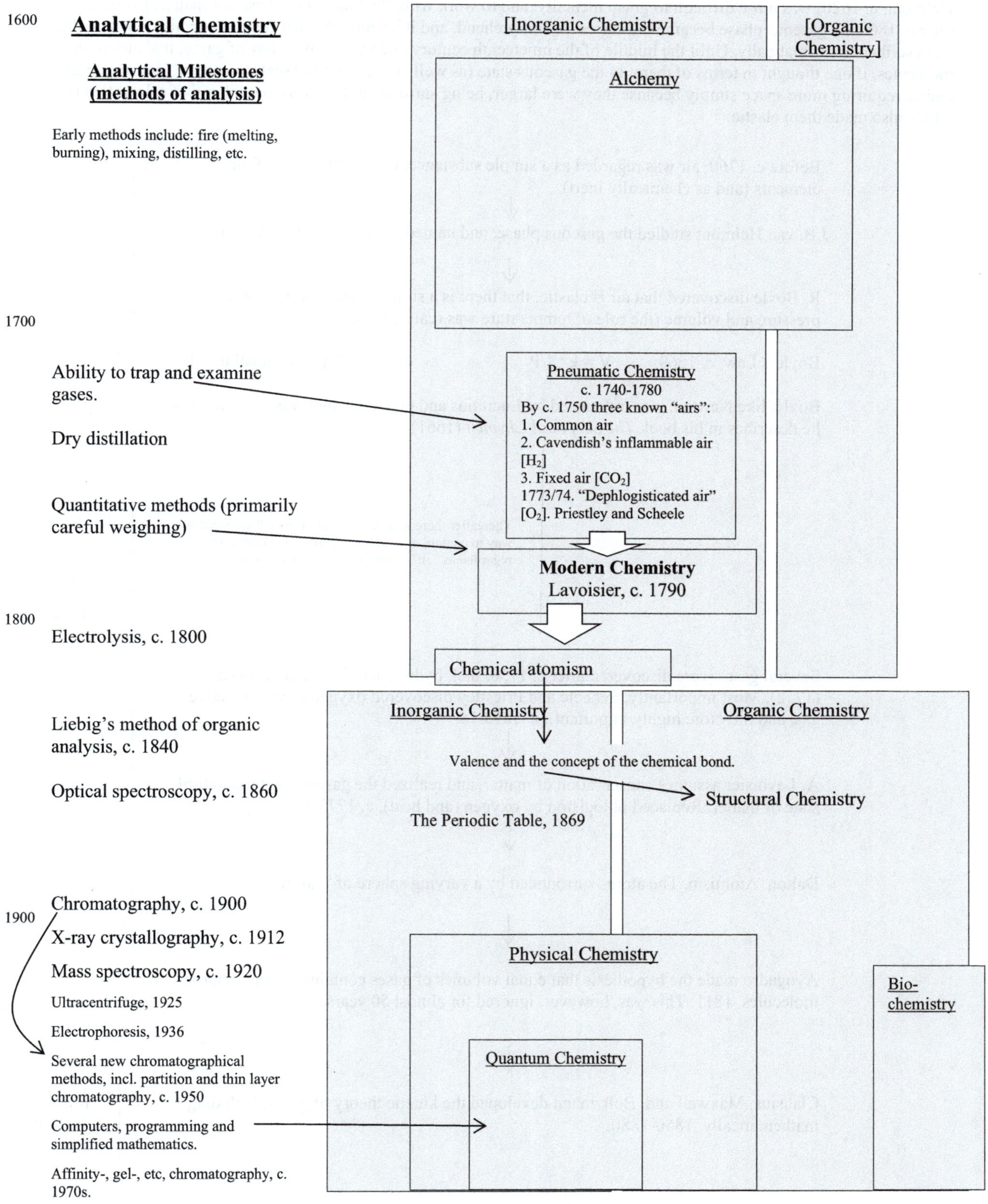

Understanding the Chemistry of Gases, c. 1670–1880: 1. Overview

The understanding of gases was fundamental for chemistry. Before that, matter seemed to disappear or emerge (e.g. oxidation) out of nowhere. At first, before c. 1750, the gaseous phase (which was named after chaos by J.B. van Helmont, c. 1620) was more difficult to grasp mentally, and to work with, than the other phases of matter. However, after c. 1800 the gaseous phase became the easiest to comprehend, and it became the first field of chemistry that could be described mathematically. Until the middle of the nineteenth century one had a static view of gases, that atoms and molecules, if one thought in terms of them, in the gaseous state (as well as the other phases) were static (non-moving), and as requiring more space simply because they were larger, being surrounded by a varying volume of caloric (heat), which also made them elastic.

Before c. 1760, air was regarded as a simple substance, one of Aristotle's four elements (and as chemically inert).

↓

J.B. van Helmont studied the gaseous phase, and named it gas (after chaos), c. 1620

↓

R. Boyle discovered that air is elastic, that there is a simple relationship between pressure and volume (the role of temperature was realized later).

Boyle's Law, c. 1660: $V = k^* 1/P$ or V proportional to $1/P$

Boyle, like Newton, was influenced by Lucretius and Gassendi, and was an atomist as he describes in his book *The Sceptical Chymist* (1661).

↓

Thereafter there was little progress since there were no way to contain or to identify gases (all gases were regarded as "air", sometimes as contaminated "air").

Several gases were discovered after c. 1750, incl. hydrogen (1766) and nitrogen (1772). Most importantly, Scheele and Priestley discovered oxygen (a very reactive gas, and therefore highly important), c. 1873/74.

↓

A. Lavoisier assumes conservation of matter, and realized the gaseous state as a third state of matter. Replaced phlogiston by oxygen (and heat), c. 1775-85

↓

Dalton. Atomism. The atoms surrounded by a varying sphere of "caloric".

↓

Avogadro made the hypothesis that equal volumes of gases contain equal number of molecules, 1811. This was, however, ignored for almost 50 years.

(Static view of gases)

↓

Clausius, Maxwell and Boltzmann developed the kinetic theory of gases, including mathematically, 1850-1880.

(Kinetic theory of gases)

Understanding the Chemistry of Gases, c. 1670–1880: 2. More Details

Static view of gases:

J.B. van Helmont studied the gaseous phase, and named it gas (after chaos), c. 1620

↓

By c. 1650, E. Torricelli and O. v. Guericke worked with airpumps and vacuum, and showed that air was material and had weight.

↓

R. Boyle continued their work and discovered that air is elastic, that there is a simple relationship between pressure and volume (the role of temperature was realized later).

Boyle's Law, c. 1660: $V = k^* 1/P$ or V proportional to 1/P

Boyle, like Newton, was influenced by Lucretius and Gassendi, and was an atomist as he describes in his book *The Sceptical Chymist* (1661).

↓ Thereafter there was little progress since there were no way to contain or to identify gases (all gases were regarded as "air", sometimes as contaminated "air").

Stephen Hales invented the pneumatic trough, c. 1730, primarily for washing gas, led to that work on gases could develop.

↓

Joseph Black distinguished ordinary air from "fixed air", [CO_2], c. 1754.

↓

A few other gases were discovered, incl. hydrogen (1766) and nitrogen (1772). Most importantly, Scheele and Priestley discovered oxygen (a very reactive gas, and therefore highly important), c. 1873/74.

↓

A. Lavoisier assumes conservation of matter, and realized the gaseous state as a third state of matter. Replaced phlogiston by oxygen (and heat), c. 1775-85.

↓

Dalton. Atomism. A static and caloric theory of gases. Explained diffusion by means of different atomic sizes.

↓

J. L. Gay-Lussac proclaimed the law of whole numbers when combining volumes of gases, 1808.

↓

This law was explained by A. Avogadro (and Ampère), with his hypothesis that equal volumes of gases contain equal number of molecules, 1811. This was, however, ignored.

↓

This was revived by S. Cannizzaro, c. 1858/60, which enabled the determination of absolute atomic and molecular weights – and the unit of mole, $6^* 10^{23}$ atoms or molecules, became important in chemistry.

↓

Kinetic theory of gases:

R. Clausius, c. 1857, developed the kinetic theory of gases.
Calculated the mean speed of molecules, and the mean free path.

↓

Maxwell and Boltzmann, c. 1860s, further developed the kinetic theory of gases, and formulated the Maxwell-Boltzmann distribution. Applied statistical ideas, and that molecules moved at varying speeds. Maxwell's law, from 1859, was the first ever statistical law in physics. Could predict macroscopic properties of gases. Gave strong support for atomic and kinetic theories.

Map of the Birth of Modern Chemistry, c. 1770–1820: 1

The "revolution" which led to modern chemistry can be regarded as having occurred in two steps; quantitative methods and the understanding of combustion (and rejection of phlogiston), c. 1790, and the idea of atomism, c. 1810.

Better understanding of gases (and the discovery of new gases, especially oxygen ("phlogiston-free air") constituted a precondition for the development of modern chemistry.

↓

A. Lavoisier and Quantitative Methods

1. Phlogiston exchanged by oxygen (and heat). Led to a new understanding of oxidation, burning and respiration.
2. Quantitative methods (careful weighting).
3. A better view of the elements, in part empirically defined, incl. giving a list of 33 elements.
4. New and better names of chemicals (compounds), which could be universally used.
5. Lavoisier's *Elements of Chemistry* (1789). Much read and facilitated the teaching of chemistry.

However, Lavoisier also retained much of "old chemical thinking" and made fundamental errors, such as including several non-elements among his list of elements (incl. heat and light), believing that oxygen was the acid-principle, regarded gases, such as oxygen, as a combination of the element and heat. He also objected to atomism (as being metaphysical)

↓

B. Atomism and Stoichiometry.

1. Atomism. The atoms of each element are identical, but different from those of other elements. This led to the determination of relative atomic weights (Dalton and Berzelius). Atomism allowed for a better understanding of chemistry, and would eventually lead to concepts such as valency and the chemical bond.
2. Stoichiometry – better understanding of the fixed proportions in chemical reactions – the laws of definite proportions and multiple proportions. Better understanding of the difference between compound and mixtures (Proust, Gay-Lussac and Avogadro).
3. New signs to signify elements and compounds (and by the 1840s a new manner of writing formulas for chemical reactions – still used today).

Map of the Birth of Modern Chemistry, c. 1770–1820: 2

Preconditions:

Better understanding of gases
↓
Discovery of oxygen, 1874

Quantitative methods

The concept of pure substance

First stage of the chemical revolution:

Lavoisier and Modern Chemistry, 1770-1800

The adoption of quantitative methods

The adoption of empiricism and positivism

Better understanding of the elements
↓
Better understanding of chemical compounds (molecules)
↓
Better understanding of chemical reactions
↓
The ancient theory of conservation of matter fully accepted

A new systematic manner of chemical terminology

Understanding of respiration
Highly important for medicine and biology.

The Laws of Stoichiometry
The regularities in the weight proportions in which elements combine with each other determined; J. Proust, J. Richter, J.L. Gay-Lussac and A. Avogadro.

Second stage of the chemical revolution:

Dalton and Chemical Atomism, 1800-1820

Atomism, which conceptually helped understanding stoichiometry and chemistry.

Better understanding of gases (although still static, and assuming the caloric theory of heat).
Explain diffusion

Better understanding of chemical reactions

First table of relative atomic weights (which eventually would lead to the periodic table)

The assumption that chemistry was connected with electricity

The development of a simple nomenclature of the elements (Berzelius)

A new better system to describe and denote chemical reactions, although generally used only after c. 1860.

Led eventually to the development of valency and the concept of the chemical bond.
↓
Structural chemistry and the possibility to grasp molecular structure.

No satisfactory explanation of chemical bonding (and thus also of reactions) could be developed before the discovery of the electron, 1897, and the development of quantum mechanics, 1925-27.

Map of the History of Modern Chemistry and Its Major Subfields, 1800–1970: 1

Chronologically Proportional

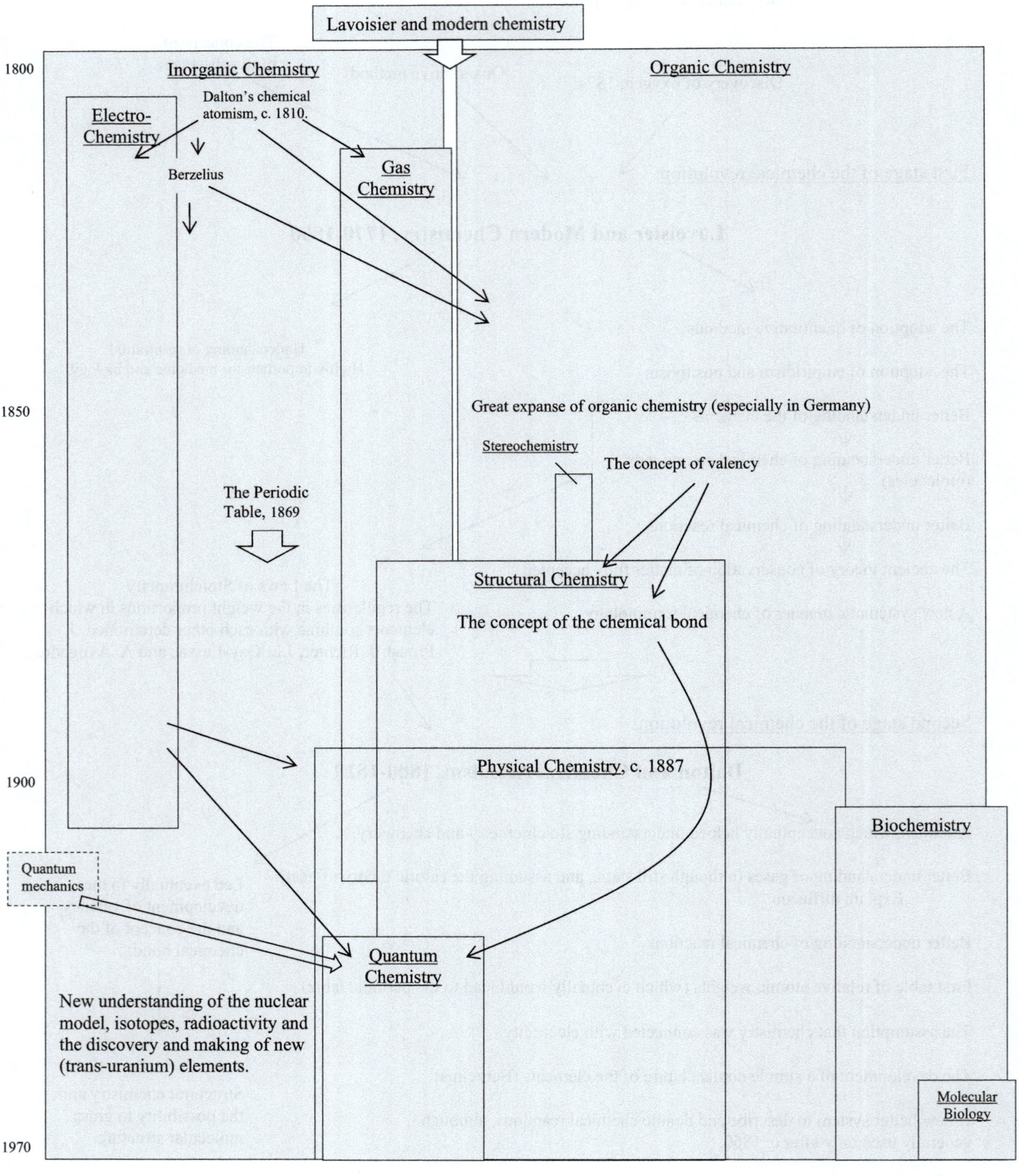

History of Modern Chemistry and Its Major Subfields, 1800–1970: 2. Names

Chronologically Proportional

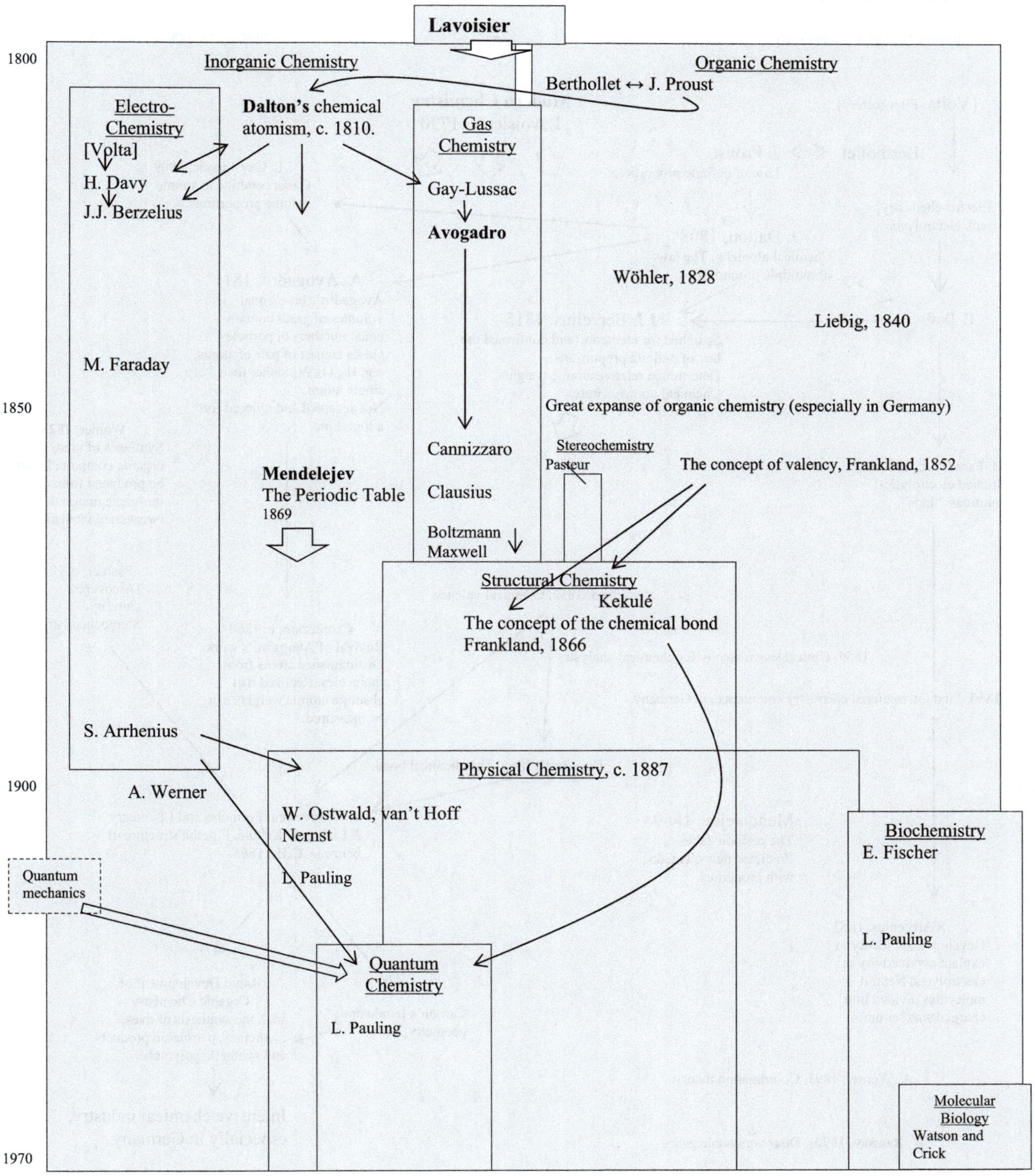

The History of Chemistry and Its Major Subfields: c. 1800–1950

The different subfields of chemistry overlap more than the subfields of physics and biology, which is not always possible to well illustrate.

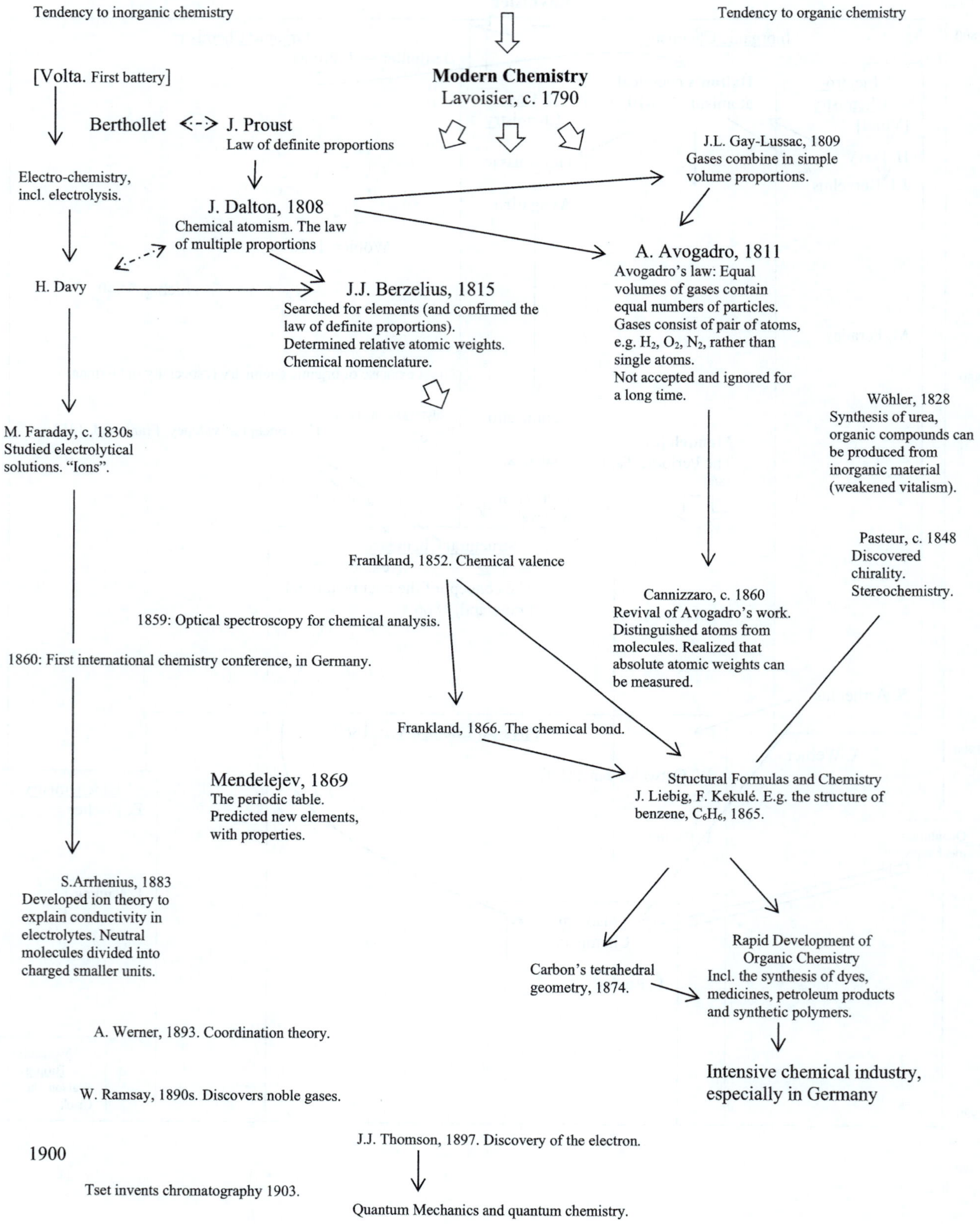

History of Physical Chemistry, c. 1800–2000

Physical chemistry applies the principles, methods and concepts of physics, such as motion, energy, thermodynamics and statistical mechanics, to chemistry. It can be regarded as having a pre-history before c. 1800 (it can, e.g. be regarded as an extension of electrochemistry and gas chemistry), but it is only with the better knowledge of the elements and chemistry during the nineteenth century that physical chemistry really began. The field grasped over both organic and inorganic chemistry, and was the first field of chemistry in which more advanced mathematical treatment was evolved.

The "remaining" ones (except phlogiston) all became part of physics

Navigation
Astronomy
Gunnary

Imponderable fluids, c. 1700-1850: heat, light, electricity, magnetism, phlogiston, aether.

The work on gases also bridged the divide between physics and chemistry

Metallurgy
Pharmacy

Physics

The science of the behavior of matter in general. Kinematics.

Experimental and mathematical.

c. 1850. The principle of conservation of energy (energetics) and entropy.

Chemistry

The science of particular kinds of matter (acids, bases, metals, etc) and the taxonomy of these.

Experimental, but basically non-mathematical. Physical chemistry will become an exception to this.

c. 1790. The principle of the conservation of matter.

Physical Chemistry

Important "Subfields"

Electrochemistry
Chemical thermodynamics
Chemical kinetics
Colloid and surface chemistry
Photochemistry
Spectroscopy
Solid state chemistry
X-ray crystallography
Quantum chemistry

> Physical chemistry was generally recognized as a subfield in 1887 with the founding of the journal *Zeitschrift für physicalische Chemie* by S. Arrhenius, W. Ostwald and J.H. van´t Hoff.

'Ionist' chemistry and the Understanding of Solutions

Arrhenius showed that molecules of electrolytes dissociated even in the absence of an electrical current.
This was supported by Ostwald, and further developed by W. Nernst.

Thermodynamics and Thermochemistry

1850s and 60s. M. Berthelot and J. Thomsen developed theories of chemical affinity as directly proportional to the heat of formation. c. 1885. New mathematical approach, based on Clausius (1854, 65), Gibbs (1875-77), Helmholtz (1882) and Duhem, that affinity is identified by free energy. Thermochemistry was also important for understanding chemical equilibrium.

The Development and Use of Spectroscopy

R. Bunsen and G. Kirchhoff developed optical spectroscopy in the 1860s, which became important for quantum mechanics and many fields of chemistry. In the first half of the nineteenth century mass spectroscopy and x-ray diffraction were developed, and many new kinds of spectroscopy in the second half.

Chemical Kinetics

Much improved understanding of which reactions occur, and how fast (and the function of catalysis) – and that this generally is done through transition states of higher energy. Earlier this was mainly done in bulk, but after WW II, with the aid of lasers (and quantum chemistry), it could also be studied on molecular level.

History of Electrochemistry

There was little understanding of chemical affinity, bonding and reactions until the twentieth century. However, already early it was suspected that it was electrical in nature (the only other known fundamental forces were gravity and magnetism), especially emphasized by Davy and forcefully by Berzelius. These theories ran into trouble when attempting to account for diatomic molecules of the same elements, e.g. H_2, as well as on other occasions. Until the discovery of the electron and the development of quantum mechanics no satisfactory explanation was available.

A. Volta, c. 1800.
The first battery, enabled electrochemistry

H. Davy, c. 1807
Determined two new elements by electrolysis, K and Na, and later others.
Suggested that chemical affinity was predominantly electrical in nature.

J.J. Berzelius
Continued and strengthened the electrical theory of chemical affinity (as well as atomism). Had a dualistic view of attraction. It could not explain binding between diatomic molecules, e.g. H_2 and O_2. Chemical reaction is the coming together of atoms containing unequal amount of positive and negative fluid.

M. Faraday
Worked on electric conduction of solutions, and formulated two laws of electrolysis (1834).
Dualism
(Named many of the terms of electrochemistry; ion, electrode, electrolysis, electrolyte, cation, anion, etc.)

R. Clausius, c. 1857
Some molecules are already ionized before current was applied to a solution.

S. Arrhensius, c. 1887
Worked on ion-chemistry. Determined that ionization of electrolytes were only complete at infinite dilution, otherwise partial ionization.

Berzelius' theory continued to influence the view and concept of chemical affinity.

Radical theory of organic chemistry.
1850s. H. Kolbe.

Frankland and the concept of valence, 1852

Frankland and the concept of chemical bond, 1866

The discovery of the electron.
J.J. Thomson, 1897. This finally confirmed that chemical affinity is electrical.

Quantum mechanics and quantum chemistry.
Understanding of chemical bonding and affinity.

Map of the History of Theoretical and Quantum Chemistry, c. 1800–2000

Only with quantum mechanics was a full understanding of chemical binding and reactivity possible, but the field also has a pre-history before that.

Atomism:
Dalton, Berzelius, *et al.*

Chemical affinity as electrostatic in nature:
Davy, Berzelius, Faraday.

The concept of valency and chemical bond:
Frankland suggests the concept of valence number in 1852. E.g. carbon having valency 4 [oxidation number].
In 1866 Frankland expands and begin to speak of chemical bond.

Structural Chemistry
Highly important development, especially for organic chemistry.

Carbon's tetrahedrical structure, c. 1874.
Kekulé, van't Hoff and A. le Bel.

Explained optical isomerism

Discovery of the electron,
J.J. Thomson, 1897.

N. Bohr, 1913
The old quantum theory

Several electrostatic models (with static electrons) ~~for bonding~~ was suggested.

G.N. Lewis.
Electrovalency, 1902
Sharing of electron pairs, 1914-1923.

Lewis-Langmuir's theory of bonding.
Ionic and covalent bonding. 1916-1920

The development of quantum mechanics,
1925-27
$H\Psi = E\Psi$

Quantum Chemistry

The first chemically relevant quantum mechanical work was done on H_2 and H_2^+.
Two different approaches were developed.

The atomic orbital (A.O.) method by W. Heitler and F. London in 1927

The molecular orbital (M.O.) approach by F. Hund.

Developed independently by Slater and Pauling. Could explain directed chemical valency by mixing s- and p- electron orbitals (therefore sometimes called the valence bond approach). Pauling also discussed the hydrogen bond.

Developed by R. Muliken, E. Hückel, C. Longuet-Higgins and C.A. Coulson.
At first the two approaches gave similar results, but the A.O. method seemed to many mathematically simpler. However, with the use of high-speed computers after WW II the M.O. method took over.

1960s. Orbital symmetry rules. R.S. Woodward and R. Hoffmann. Allowed specific predictions of stereo-chemical details and reaction outcomes.

With faster and cheaper computers, and better approximation methods, quantum chemistry was able to help understand and predict many chemical properties and problems, from protein folding to the synthesis of medicines.

History of Inorganic Chemistry: Overview

Inorganic chemistry is the study of the elements and their compounds, except those of carbon. This includes metallurgy and the properties of many inorganic dyes, as well as aspects of glass-making and ceramics, all of which has a long history. Only in the eighteenth century did one begin to explicitly separate inorganic from organic chemistry. Until the first half of the nineteenth century the history of inorganic chemistry overlap with the history of chemistry generally – most of the discoveries and theories behind the chemical revolution, c. 1780-1820, were done using inorganic substances.

New elements found, and atomic weights determined

Unlike the case for organic chemistry after c. 1860, there were no good structural theory and understanding of inorganic structures.

The periodic table, 1869
This led to better understanding of the elements, and that new elements were found.

A. Werner, c. 1893
Revolutionized **coordination theory**. A central (metal) atom surrounded by a definite number (coordination number) of other atoms or groups

Much work was also done on metal alloys.

The discovery of the noble gases in the 1890s, and then included into the period table.

The work on separating uranium isotopes in the 1940s (for the Manhattan project) led to a revival of many fields of inorganic chemistry.

Inorganic chemistry has greatly developed during the twentieth century, both in regard of understanding the processes and reactions (including catalysis) and of material and industrial production.

Material science since c. 1960 is largely based on advances in inorganic chemistry.

History of Organic Chemistry: Overview

Organic chemistry has a much longer history, but it can nonetheless be reasonable here to begin c. 1800 with Lavoisier's discoveries and Berzelius' explicit distinction between organic and inorganic chemistry (1807) – at a time when the theoretical understanding behind chemistry was getting much more detailed than earlier.

A. Lavoisier, c. 1790
Showed that most organic compounds only contain carbon, hydrogen and oxygen (and sometimes nitrogen).

G.L. Gay-Lussac, 1811
Constructed more detailed empirical formulas for simple organic molecules.

Work on organic macromolecules.
Lipids, proteins and carbohydrates (sugars).

Carbohydrates could be reduced to sugars (glukos).
A few amino acids were identified as part of the larger proteins.

J.J. Berzelius, 1807
The most influential chemist during the first half of the century. Distinguished between the chemistry of living matter (organic) and the rest (inorganic).

F. Wöhler, 1828
Synthesis of urea (organic substance from inorganic sources).

J.v. Liebig
Very influential German organic chemist and teacher. Simplified organic analysis. Studied isomers, substances with the same empirical formulas, but different properties.

Dumas, c.1830
Influential French organic chemist.

L. Pasteur
Stereomers, c. 1848

"Radical" Theory of Organic Chemistry

Organic compounds could be described as constructed from smaller groups of atoms (radical), e.g. CN, etc.

H. Kolbe, 1845
Synthesized several organic acids, and thus strengthened Wöhler's work.

Type theory of organic chemistry.

Dumas. 1830-60.

M. Berthelot, 1850
Synthesized many organic substances (but all very small). He also worked on and synthesized a few large organic lipids.

E. Frankland, 1852
Valence. Each sort of atom can bind only a definite number of others.

F.A. Kekulé, 1857 and the 1860s
Defined organic chemistry as chemistry dealing with carbon-containing molecules, 1861. Developed the concept of valence and structural formulas. This concept of structures with connected atoms won over the "radical" concept. It was now possible to understand isomers. There was no successful formula for benzene, C_6H_6, until Kekulé 1865. But these structures were still in two dimensions.

J.H. van't Hoff, J.A. Le Bel, 1874.
Carbon's tetrahedral structure.

This led to greatly improved understanding of organic structural chemistry, and thereafter to a rapid development of organic chemistry and industrial chemistry.

Physical chemistry → Better understanding of organic reaction mechanisms and kinetics.

Quantum chemistry → Better understanding of bonding, the hydrogen bond and L. Pauling was able to explain the unusual properties of benzene.

New analytical methods and instruments → In the period 1930-1960 many new analytical instruments developed, e.g. pH-meters, mass spectroscopy, gas chromatography, NMR spectroscopy, etc.

History of Biochemistry: Overview

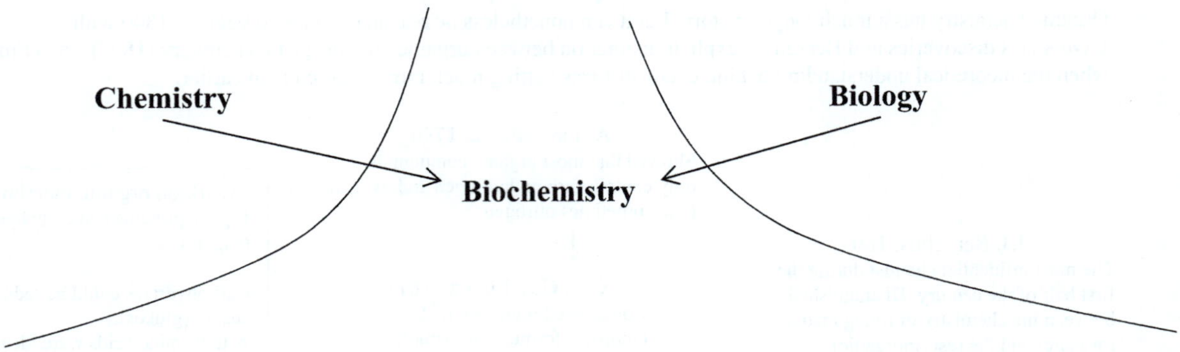

The study of the chemical constituents and processes of living systems was born at the end of the 19th century out of organic chemistry, but also from cell biology and physiological chemistry. However, precursors to it existed earlier under the name iatrochemistry, animal and plant chemistry and/or physiological chemistry. Lavoisier showed at the end of the 18th century that animal substances contain four elements, carbon, oxygen, nitrogen and hydrogen, and showed that alcoholic fermentation (the first biologically important process chemically studied) constituted of conversion of sugars to alcohol and carbonic acid. Together with Laplace, he also showed that respiration is slow combustion. By 1840 one had amassed much information about the chemistry of living systems. J.J. Liebig combined much of this information with the concept of force in his *Animal Chemistry* (1842) to produce a dynamical system revolutionizing biologists' understanding of plant and animal physiology. Further work, inspired by Liebig, or by opposition to it, led to new information about several fundamental aspects of biochemistry.

The term 'biochemistry' was coined by F. Hoppe-Seyler (who also was the first to isolate nucleic acid in 1869) in the first issue of his periodical, *Zeitschrift für physiologische Chemie* (1877), but the field was not fully founded and institutionalized until early in 1900. The first professor in biochemistry was appointed in 1914.

Fermentation

Knowledge of alcoholic fermentation is ancient. Lavoisier and others assumed that it was a purely chemical process. By 1850 this was opposed by cell biology which showed that yeast globules were involved, culminating in the Liebig-Pasteur debate. Pasteur, who argued that whole yeast cells were involved, won, c. 1860. However, Buchner produced fermentation in cell-free extracts in 1897, and this was an important step to understanding enzyme chemistry and is sometimes regarded as the beginning of biochemistry.

Proteins and Enzymes

Proteins were named in 1838, but the work on them developed slowly during the 19th century, although already in 1836 pepsin, the first enzyme, was identified (named in 1876). Early in the 20th century it was discovered that proteins are built by amino acids, linked with peptide bonds (-CO-NH-), but their sizes and structure remained uncertain. Not until x-ray crystallography at the beginning of the 1930s, was progress possible. L. Pauling, between 1932-52 proposed the "alpha-helix" model, confirmed in the 1960s.

The study of enzyme reactions occupied much of the first half of the 20th century. The complete sequence of amino acids in insulin between 1945-55.

Chemical Cycles: Metabolism

Between 1900 and 1940 the central processes of metabolism, both anabolic (synthesis) and catabolic (breakdown) pathways, were described.

By 1937 the "citric-acid" (Krebs) cycle, the final common pathway for the decomposition of all classes of foodstuff, was discovered, and about the same time the ATP-cycle, yielding the energy of cells.

History of Molecular Biology

From c. 1960, the chemical processes related to heredity (DNA) and the synthesis of proteins and nucleic acids had become an independent field of scientific inquiry.

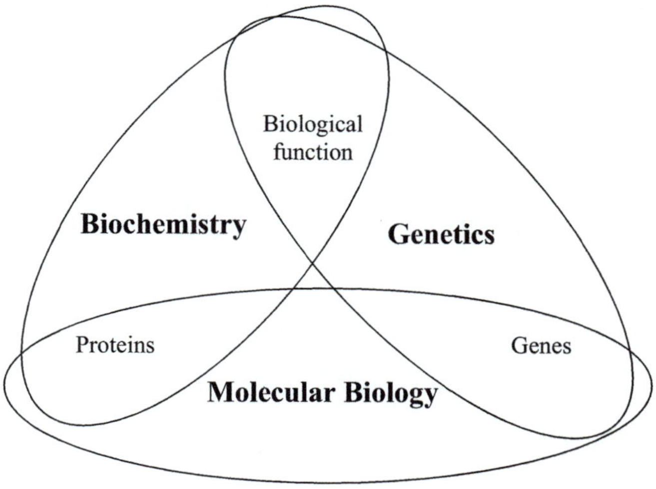

O.T. Avery *et al*, as part of the Delbrück phage group, showed in 1944 that nucleic acid carries the genetic information – previously assumed to be done by proteins.

G. Beadle, 1945, One gene, one enzyme-hypothesis.

Linus Pauling ←-·-·-·→

1953. DNA. Double helix structure. R. Franklin. Watson and Crick.

1961. F. Jacob and J. Monod showed that some genes can regulate the expression of other genes. Hypothesized the existence of mRNA, found soon afterwards.

↘ 1960-65. Discovery of RNA, mRNA and tRNA.

↘ 1964. The genetic code: three base pairs in the DNA and RNA code for specific amino acids.

↘ Increased complexity of gene activity (knowing the genetic code not enough) – can be context-dependent etc.

↘ Genome projects, inclusive HUGO (2001).

6 The Development of the Exact Sciences during Different Historical Epochs: 1500–2020

T. H. Brobjer, *Atlas of the History of Modern Science 1500-2020*, https://doi.org/10.1007/978-3-031-76627-5_6

History of Science 1500–2000: Overview

Chronologically Proportional

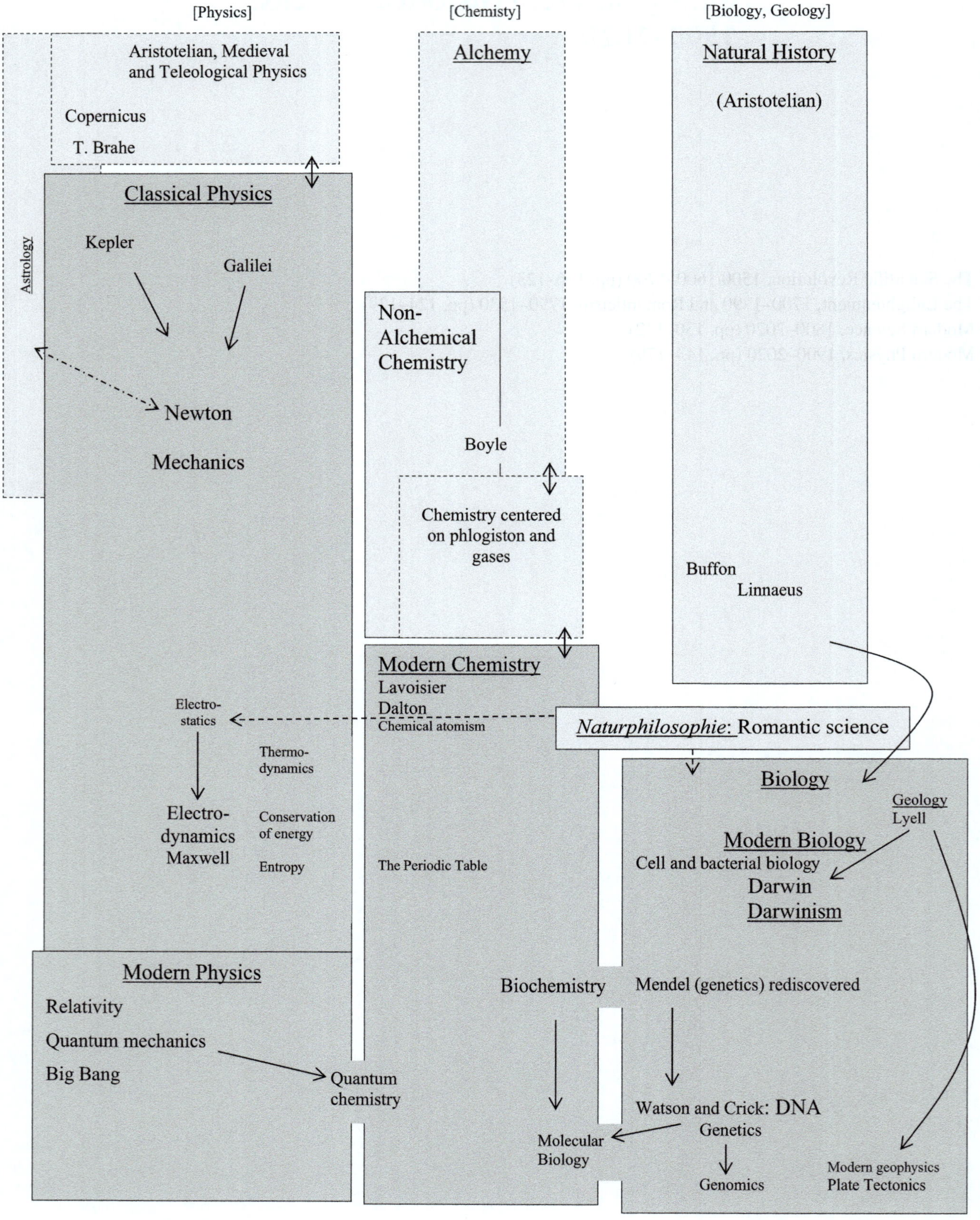

Background: 1. Philosophers of Nature during the Middle Ages and the Renaissance

The Middle Age, 400–1400

Early Middle Age
Preservers and popularizers of Greek and Roman knowledge:
Boethius, 510 ("the last Roman and the first scholastic"). Logic. Translated two logical works by Aristotle.
Cassiodorus, 530
Isodore of Seville, 600

The Carolingian mini-Renaissance of antiquity c. 780-887. Alcuin, 780
Johannes Scotus Erigena, 850

New ancient texts via Arabic and Greek, c. 1150-1270. Especially Aristotle.
Earliest universities, c. 1200 (almost without exception with Aristotelian curriculum): Paris, Bologna, Oxford, etc.

Aristotelians	Critical Aristotelians	Platonists	Atomist/Skeptic

Scholasticism (based on newly available Aristotelian logic, the complete *Organon*, the so-called "*logica nova*")

Nominalism	↔	Realism (universalia)	
(Aristotelians)		Robert Grosseteste, 1210	
Albertus Magnus, 1240			
Thomas ab Aquino, 1260			
	Roger Bacon, 1270		
	Duns Scotus, 1300		
	William of Ockham, 1320		
	Jean Buridan, 1330		
	Nicole Oresme, 1360		Nicolas of Autrecourt, 1340

Renaissance "Scientific" Thinkers 1350–1600

Most of the more well-known Renaissance thinkers were more humanists (i.e. setting rhetoric above philosophy) than philosophers, and most of them – with a few exceptions – had little interest in natural science.

Philosophy	Science	Technology, Architecture and Art related to Science
Nicholas of Cusa, 1440		
		Gutenberg, 1450
		Alberti, 1440
	(Leonardo da Vinci)	Leonardo da Vinci, 1490
		Michelangelo, 1515
	Paracelsus, 1530	
	Cardano, 1535	
	Copernicus (1543)	
	Vesalius (1543)	
Telesio, 1563		
Patrizi, 1570		
	T. Brahe, 1585	
G. Bruno, †1600		
Campanella, 1610		
	Kepler, 1610	

Background: 2. With Summaries of Philosophers of Nature

Aristotelianism | Platonism and Neoplatonism

Preservers and popularizers of Greek and Roman knowledge:

Boethius, 500. "The last Roman and the first scholastic". Planned to translate all of Plato and Aristotle into Latin, but only translated two minor works by the latter (*Categoried* and *De Interpretatione* as well as Porphyros' *Introduction* to Aristotles' *Categories*) before executed. Even these works were not really read until after 1000.

Cassiodorus, 530. Compiled much of the available knowledge of antiquity. Used and described the seven liberal arts.

Isodore of Seville, 600 ("the last scholar of the ancient world"). Attempted to summarize all knowledge in his book *Ethymologies*, one of the most read books in the medieval period.

Charlemagne, 800, mini-Renaissance of antiquity. Education and learning organized by Alcuin, 780.

Johannes Scotus Erigena, 850. Learned Neoplatonic Irish monk who knew Greek and translated many works by the Church Fathers, incl. Dionysius the Areopagite's Neoplatonic works into Latin. Influenced later mystics.

Anselm of Canterbury, 1070
(Defended the importance of logic/reason and that it was compatible with faith, but the latter as primary ("faith seeking understanding"). Produced an influential ontological proof of the existence of God.)

Abelard, 1120 ↔ **Bernhard of Clairvaux**, 1130 (Opponent of Abelard and the new logic.)
(Logician, inspired by Boethius' translations of Aristotle's logic, begins the realism/nominalism debate. Himself a nominalist. Strong believer in the importance of logic. Argued that man is free and responsible. Wrote *Sic et Non*.

Scholasticism (based on Aristotelian logic)

Albertus Magnus (Aristotle scholar), 1240. Wanted to make all knowledge from antiquity, especially Aristotle, known.

Thomas ab Aquino, 1260
(Made a synthesis of Aristotle and Christianity. The most influential thinker in the medieval period. Created medieval philosophy and re-created medieval theology.)

Roger Bacon, 1270 (early "empirist", for experiment and observation.)
Duns Scotus, 1300.
William of Ockham, 1320 (Ockham's razor, faith / reason,
Jean Buridan, 1330 (impetus)

Nicolas of Autrecourt, 1340 (Early skeptic and atomist, who was forced by the Church to retract his statements.)

Nicole Oresme, 1360 (astronomy, all motion relative, earth may be rotating.)

Nicholaus of Cusa, 1440 (astronomer, infinite universe)

The Renaissance, c. 1350-1600, is mostly associated with the re-birth of antiquity, especially the Greeks and Plato, with developments in the arts and architecture, and with a new view of man and the liberty of man, humanism. Most humanists were not particularly interested in science. However, the development of, for example, the central perspective requires a mathematical and scientific approach, and this is equally true for much of sculpture and architecture as well (cf. Leonardo and Michelangelo's work). If we except the period after the middle of the 16th century (often treated together with the scientific revolution) there are relatively few scientific advancements, especially in a more theoretical manner.

Epochs in Modern Intellectual History: Main Characteristics Relevant for Science

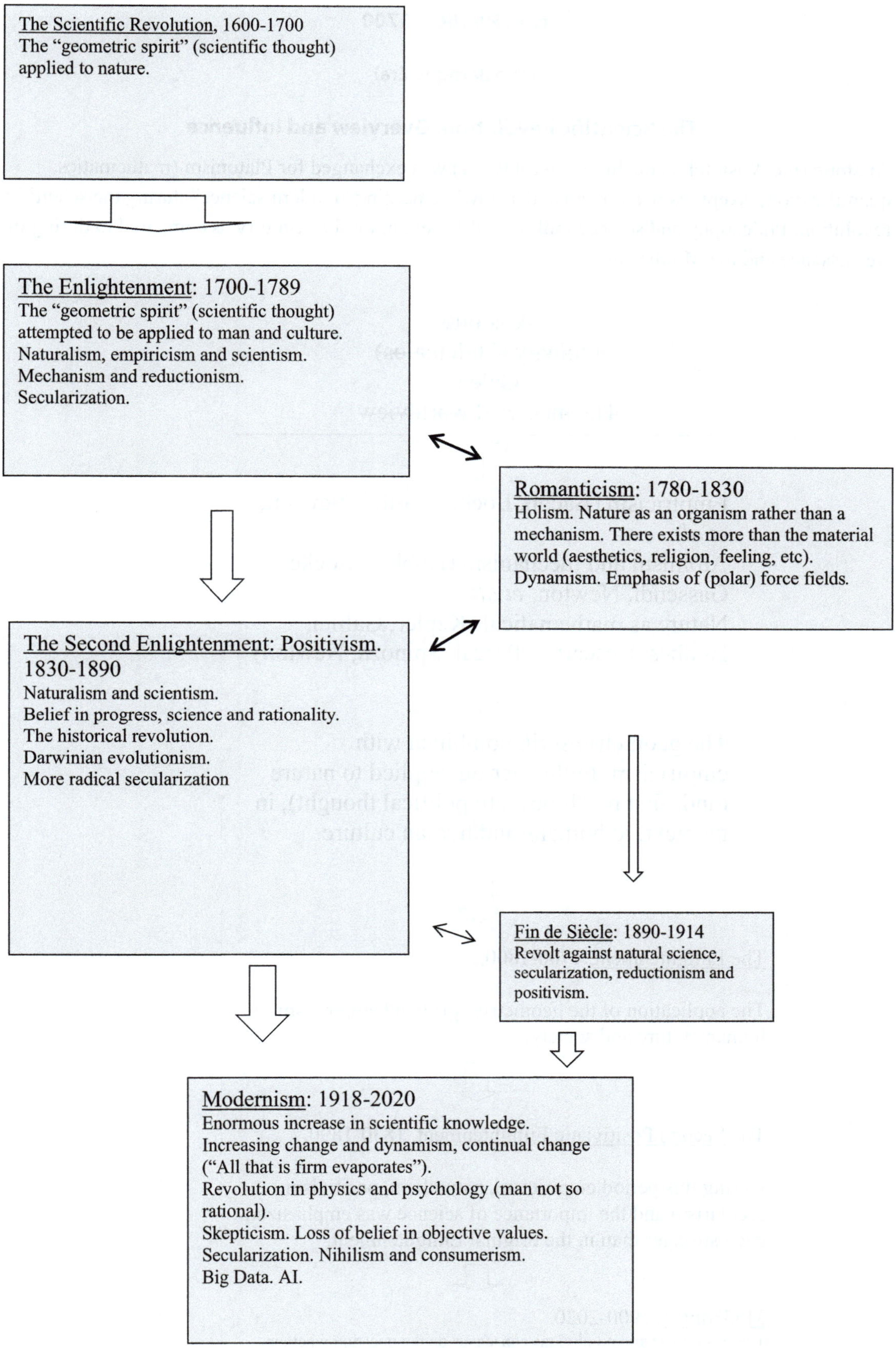

Science and Scientific Thought during the Scientific Revolution

c. 1550/1600–1700

(The Baroque Era)

The Scientific Revolution: Overview and Influence

Aristotle (the Aristotelian qualitative worldview) was exchanged for Platonism (mathematics, quantification), skepticism, atomism and a slowly emerging "modern science" during the scientific revolution. Philosophy and science look very different in the 17th century as compared to during the Renaissance and the Middle Ages.

~~Aristotle~~
~~Ptolemy (Ptolemaios)~~
~~Galen~~
~~The medieval worldview~~

⇩

Empiricism (Bacon, Locke, Galilei, Newton, *et al*)
Atomism and mechanism (Hobbes, Locke, Gassendi, Newton, *et al*)
Nature as mathematical (Kepler, Galilei, Hobbes, Descartes, Pascal, Spinoza, Newton)

⇩

The geometric sprit, combined with empiricism. In this period, applied to nature (and, first by Hobbes, to political thought), in the next, to humans and human culture.

⇩

<u>The Enlightenment</u>, 1700-1800.

The application of the geometric spirit and empiricism to human culture and society.

⇩

<u>The Second Positivistic Enlightenment</u>, 1830-1890.

During this period empiricism, naturalism, positivism, secularism and the importance of science was emphasized even stronger than in the original Enlightenment.

⇩

<u>Modernism</u>, 1900-2020.

There is debate and no obvious consensus on the question whether modernism, on the whole, should be regarded as a continuation of this "scientific" and "scientistic" movement. However, there is no doubt that this tradition at the very least is a strong tendency within and contribution to modernism.

Overview of the Scientific Revolution: 1. Scientists

1600–1700, 1600–1688, 1543–1700, 1500–1700

Pre-Scientific Revolution: Aristotelian worldview, teleological, geocentric, organic, qualitative.

The transitory period, 1500-1600, was characterized by: humanism, book printing, the discovery of the New World, reforms of and schism within Christianity, the belief that everything is connected and related – *scala naturae*, "the great chain of being", the relation microcosm-macrocosm, *magia naturalis*, astrology and alchemy.

During the Scientific Revolution much of this changes to a worldview which was: non-teleological, heliocentric, mechanistic, qualitative and reductionistic. Observation and experiment slowly become more important than Christian and ancient authority.

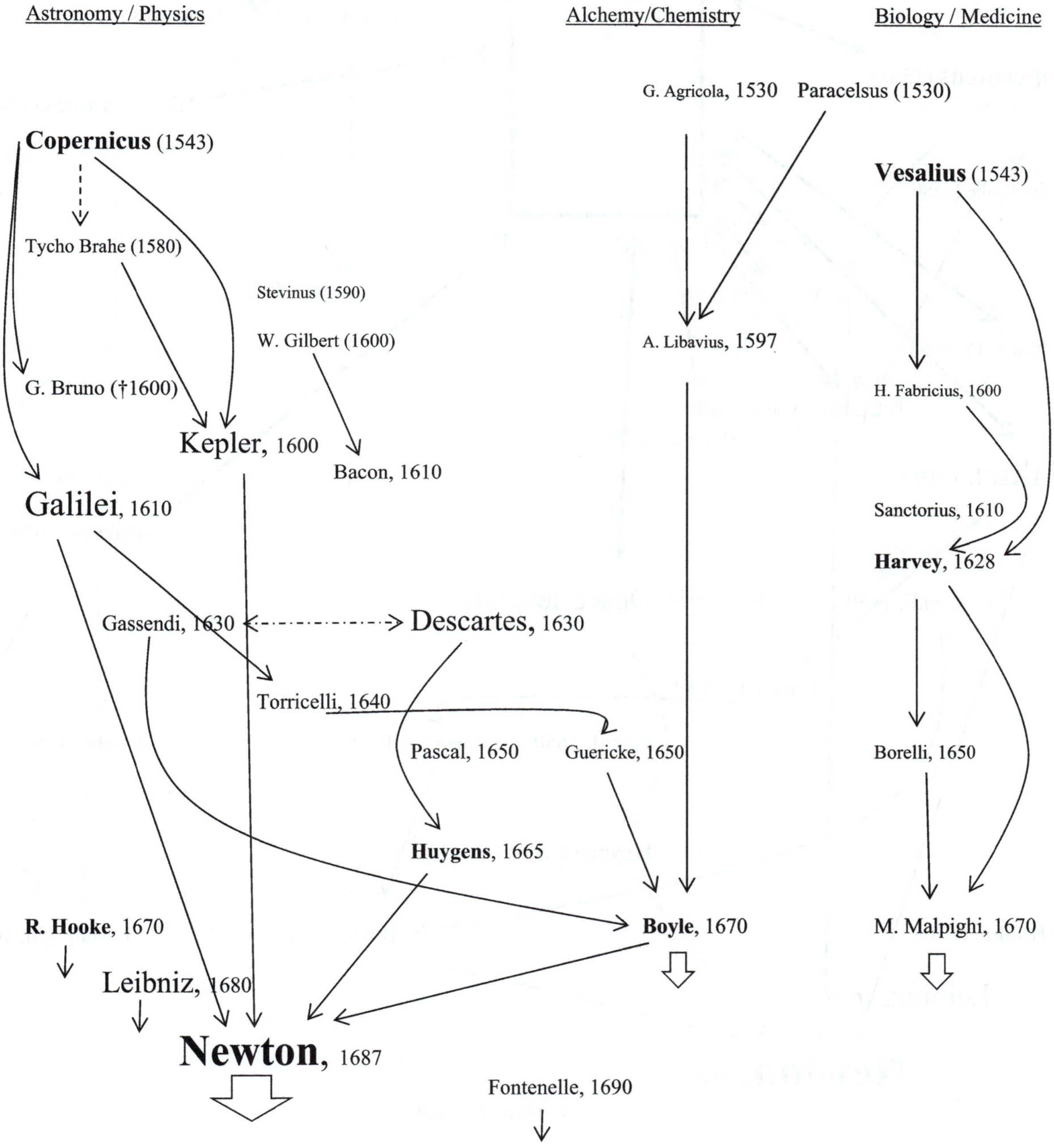

Overview of the Scientific Revolution: 2. Scientists' Opposition to Aristotelianism

1500–1700

Aristotle's natural philosophy came to constitute the center-place of the world view of the Middle Ages (except for faith and the Bible). Aristotle himself had done extensive empirical examinations, but by the Middle Ages his philosophy had become dogma and could not be questioned. However, during the latter parts of the seventeenth century more and more thinkers reached results and conclusions which did not agree with Aristotelianism, and the critique of it became one of the unifying themes of the scientific revolution.

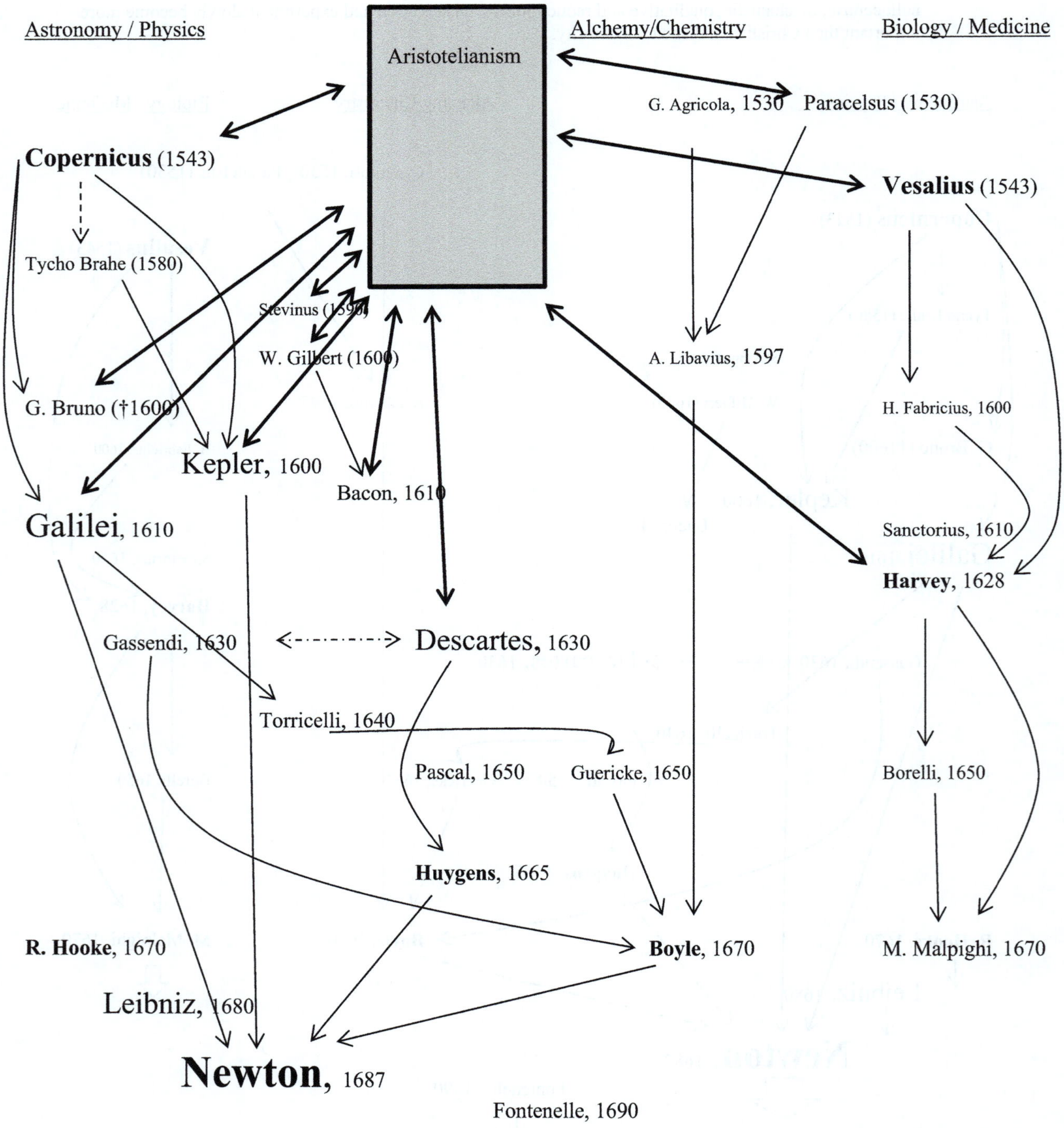

The Ancient Greek Influence on the Thinkers of the Scientific Revolution

Ancient Greek thought continued to influence and inspire much of seventeenth century thought, even to the degree that at the end of the century there was a great quarrel – the battle of the books – about whether the moderns can or could be better or equal to the ancients in any sense. The quarrel ended inconclusively, with no obvious victor. The influence of the Greeks were particularly prominent in regard to mathematics and the quantitative methods, in the growth of skepticism and the effect of Epicurean atomism, which in many ways foreshadowed much of the mechanistic worldview.

The important negative influence of Aristotelianism is not included in this map since it has been summarized above, and also because it was less Aristotle and more a dogmatic late Aristotelianism which many of the thinkers of the scientific revolution reacted against.

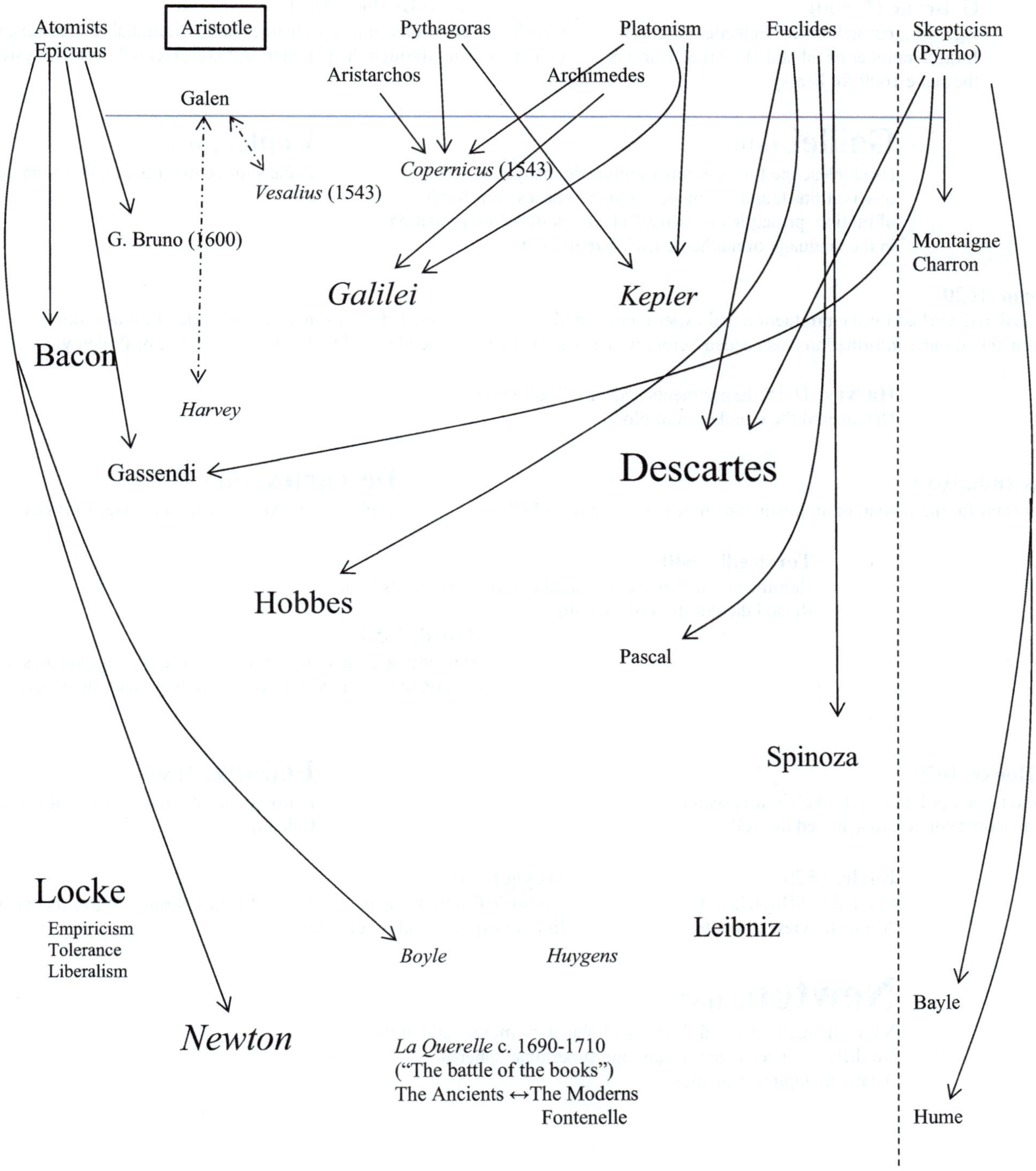

Overview of the Scientific Revolution: 2. Scientists with Summaries

Paracelsus (1530): New view of medicine: A chymical or alchemical worldview

Copernicus (1543): Platonism and heliocentric worldview

Vesalius (1543): Detailed map and concept of human anatomy

W. **Gilbert** (1600): Magnetism. The earth a great magnet. "Magnetic philosophy". Made experiments. Gravity = magnetism. Proposed a geoheliocentric cosmology, similar to Brahe's.

G. **Bruno** (†1600)
Against Aristotelianism. Eclectic. Pantheist. Infinite number of inhabited worlds. Burnt at the stake 1600 for heresy.

Tycho **Brahe** (1580):
Careful observations, and saw changes in the 'immutable' heavens, and comets passing through the planetary spheres. Geoheliocentric system.

Galilei, 1610
Used telescope for astronomy and made many new observations (Jupiter's moons, Venus' phases etc). Study of motion, principle of inertia. "The book of nature is written in the language of mathematics". Experiments.

Kepler, 1600
Planets move in ellipses rather than circles

Bacon, 1620
Argued and worked for the production and experimental methods of new knowledge. Atomist, and regarded heat as motion. Claimed that more information was needed before we can decide between the views of Copernicus and those of Ptolemy.

Harvey, 1628: Experiments and careful observations. Discovered the circulation of blood.

Gassendi, 1630
Important for the revival of atomism and "mechanical philosophy".

Descartes, 1630
Mathematics. Mechanistic universe. Vortices.

Torricelli, 1640
Barometer. Air has weight (unlike in Aristotle's system). Raised the question of vacuum.

Pascal, 1650
Mathematician and physicist, who after a religious mystical experience occupied himself exclusively with theology.

R. **Hooke**, 1670
Physicist and polymath. Hooke's law (elasticity). Microscopic studies, discovered the cell.

Leibniz, 1680
Philosopher, mathematician and polymath. Calculus.

Boyle, 1670
Studied air: Boyle's law. Atomistic view of matter.

Huygens, 1665
Versatile Dutch physicist. Discovered Saturn's rings. Wave theory of light to explain double refraction.

Newton, 1687
Mathematically derived the laws of planetary motion and gravity. No difference between the sub- and superlunar spheres. Atomism. Optics. Calculus.

Overview of the Scientific Revolution: 1. Philosophers and *Scientists*

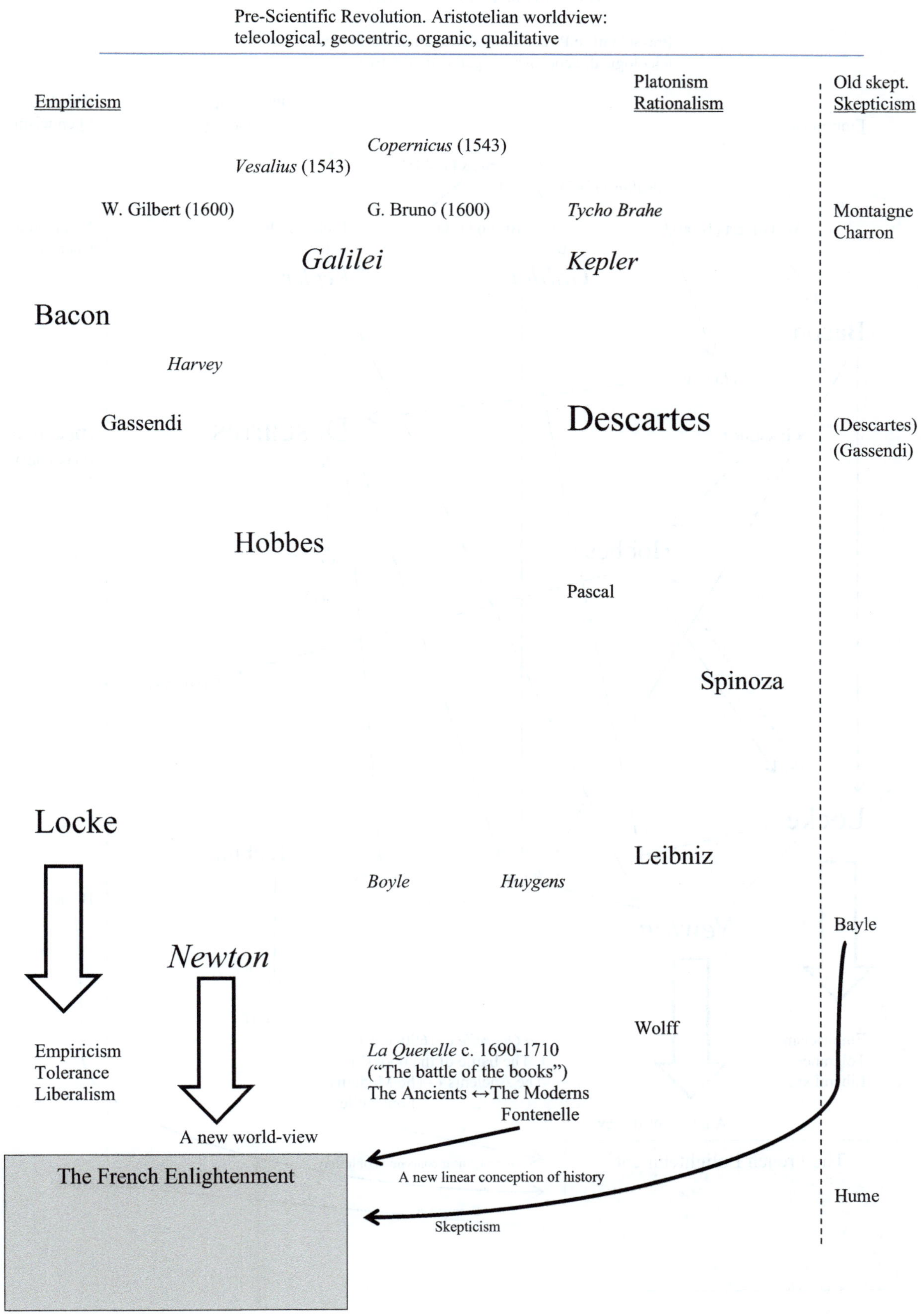

Overview of the Scientific Revolution: 2. With Relations

1600–1700, 1600–1688, 1543–1700

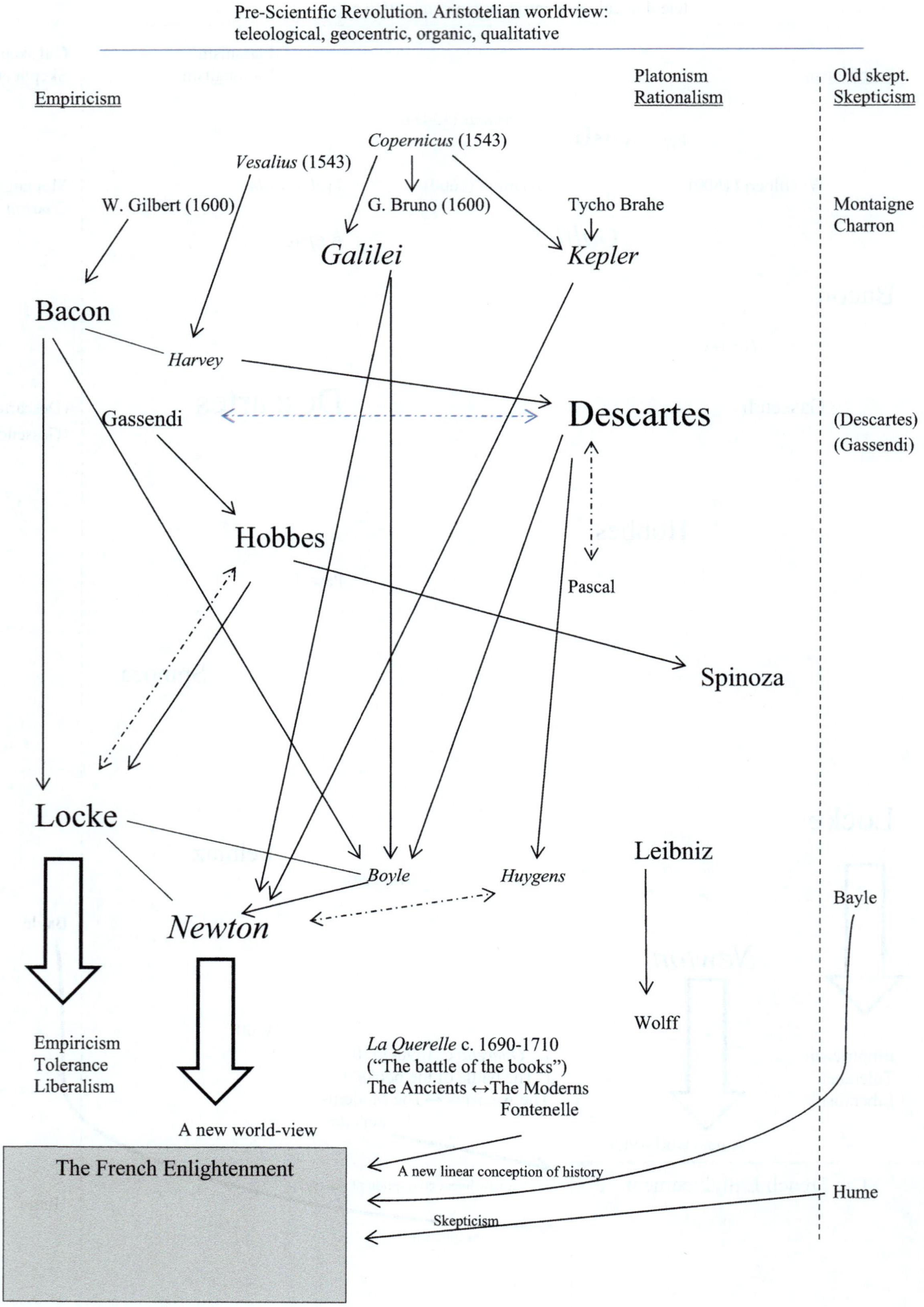

Some Prominent 17th Century Examples of the Importance of New Instruments for the Development of Science

All five of the instruments listed here were immensely important for improving man's ability to observe and to do experiments. All of them led to enormous extension of man's knowledge. They were integral parts of the scientific revolution.

The telescope (astronomical use of) Galilei, 1609	⟶	Many important new astronomical discoveries, leading to a new worldview and helping to develop physics.
The thermometer Galilei, Santorio, *et al.* c. 1615 (There had been some use of thermometers in antiquity)	⟶	An important instrument for the ability to make quantitative observations and experiments, especially in physiology.
The microscope Invented in Holland (like the telescope), c. 1590, but the first published works based on the microscope was in 1625 and 1665.	⟶	Showed man that there were many things on the microscopic level than man earlier had not known (not even the ancients). Led to the discovery of cells etc. 1625 *Melissographia*, microscopic study of a bee. 1665 Hooke *Micrographia.*
The barometer and the air pump The first water barometer, G. Barti, 1641. Improved mercury barometer by the Galilei student Torricelli 1644. The first use of an air pump to create vacuum was Otto von Guericke 1647, and later by Boyle.	⟶	Important for the understanding of air, gases and of vacuum, which aided chemistry to develop. Later (around 1900) also important for the discovery and understanding of x-rays and beta-rays (electrons).
The pendulum Galilei discovered that the time for each oscillation was nearly independent of the length of the pendulum (isochronous). Ch. Huygens used pendulum to control clocks, and made important theoretical studies of the pendulum (1657, 1673). Jean Richter and Newton noticed that the pendulum can be used to measure gravity (and the shape of the earth), 1672.	⟶	This gave possibility to detailed time and gravity (geodesy) measurements, as well as improved navigation. The pendulum led to improved clocks (earlier clocks used weights) and by 1728 to John Harrison's stable clocks which enabled the determination of longitude at sea and better navigation.

Copernicus (1473–1543)

1. The simple picture:

Geocentric worldview —(Copernicus, 1543)→ Heliocentric worldview

---

2. A somewhat more complex and accurate picture:

The generally accepted view, c. 1500
Geocentric worldview, based on Aristotle and the Ptolemaic system, with spheres, epicycles and eccentric cycles. Some 80 cycles needed, and for some of them the movement was not constant (in regard to the centre).

↓

Copernicus, as a Platonist, found this system too complicated and asymmetric, and wanted to simplify it – especially to do away with the eccentric aspects of the cycles.
Learnt from antiquity about attempts with "the absurd idea" of not having the earth at the centre.
Copernicus contined to assume that heavenly bodies move in perfect circles with constant speed.

↓

This led to a heliocentric system, but still with spheres (all stars equally distant from the sun) and epicycles (and less accurate than the Ptolemaic system, due to inaccurate measurements and the fact that planets actually move in ellipses), but simpler and conceptually easier. 48 cycles was still need in Copernicus' system.

In some ways this was just half a "revolution".

Consequences:

1. A new heliocentric worldview (although it took almost a century before it was accepted, and only after Kepler was there definite proofs that a heliocentric model was better).

2. It would eventually lead to a severe dispute with the Church.

3. The Copernican model was not very accurate. New observations were necessary.
→ T. Brahe → Kepler

4. A new physics was needed to explain falling bodies and the motions of the planets.
(Dropped things do not fall towards the centre of the universe.)
(Why do we not notice the fast circulation and rotation of the earth?)
→ Galilei → Newton

5. Earth's special place – not only at the centre, but as very different from the other planets – was denied.

6. There was no need of a Prime Mover (placed beyond the sphere of the stars) any longer – the sun ruled from the centre.

From a Geocentric to a Heliocentric Worldview: Comparison of Aristarchos with Copernicus

Both of them assumed the Pythagorean-Platonic idea that perfect motions, and thus also astronomical movements, were circular, constant and eternal.

	Aristarchos (c. 310 - c. 230 BC)	Copernicus (1473-1543)
Education	Little is known. Studied under Strato, third leader of Lykeion, and possibly also in the Academy.	Received a full humanist education, at several European universities.
Context	General belief about the sky: Concentric spheres, immaterial (mathematical) or physical (aether). Probably a fairly small spherical universe, with all the stars at the same distance from the earth.	General belief about the sky: A combination of two theories: Eccentric circles and epicycles combined with a belief in concentric physical (aether) spheres.
Proposals	He proposed a heliocentric system, with the earth rotating and circulating around the sun, in a very large universe. Wrote a book about this, but only his other book survives: *On the Sizes and Distances of the Sun and the Moon*. Since there is no parallax (shift of the stars between spring and autumn), he assumed that the stars must be very far away. His universe seems to have been about four light-years in radius.	He proposed a heliocentric system, with the earth rotating and circulating around the sun. *On the Revolutions of the Celestrial Orbs* (1543) His arguments were mostly mathematical. He wanted to improve Ptolemy's system, to make it more Platonic and Pythagorean. He was able to reduce the number of circles needed from about 80 for the elaborate versions of the Ptolemaic system to about half that number. Copernicus system was actually mostly less accurate than Ptolemy's. The probably most revolutionary of his thought was that it removed the 'unmoved mover' (of Aristotle), and instead placed the sun as the central force.
Influence	Almost no known direct influence, except on the astronomer Seleucus (c. 150 BC), but possibly influenced Copernicus much later. His theory was rejected by Greek religion and general conservatism. Also by contemporary physics, 'gravity' would not work. We ought to notice if the earth moved and rotated. Almost 'proved wrong' by the fact that there was no parallax.	Slow and little influence for the first one or two generations. Opposed by the Church and astrology. Accepted by Galilei, near 1610, and by many by the middle of the seventeenth century.

The Revival of Epicurean Atomism during the Seventeenth Century: 1. Overview

At the beginning of the century, two different forms of element teachings, the four-element scholastic Aristotelianism and the three-principle system of Paracelsus, ruled in regard to the nature of matter. By the latter parts of the century, atomism had taken over the place as the most acceptable theory of matter.

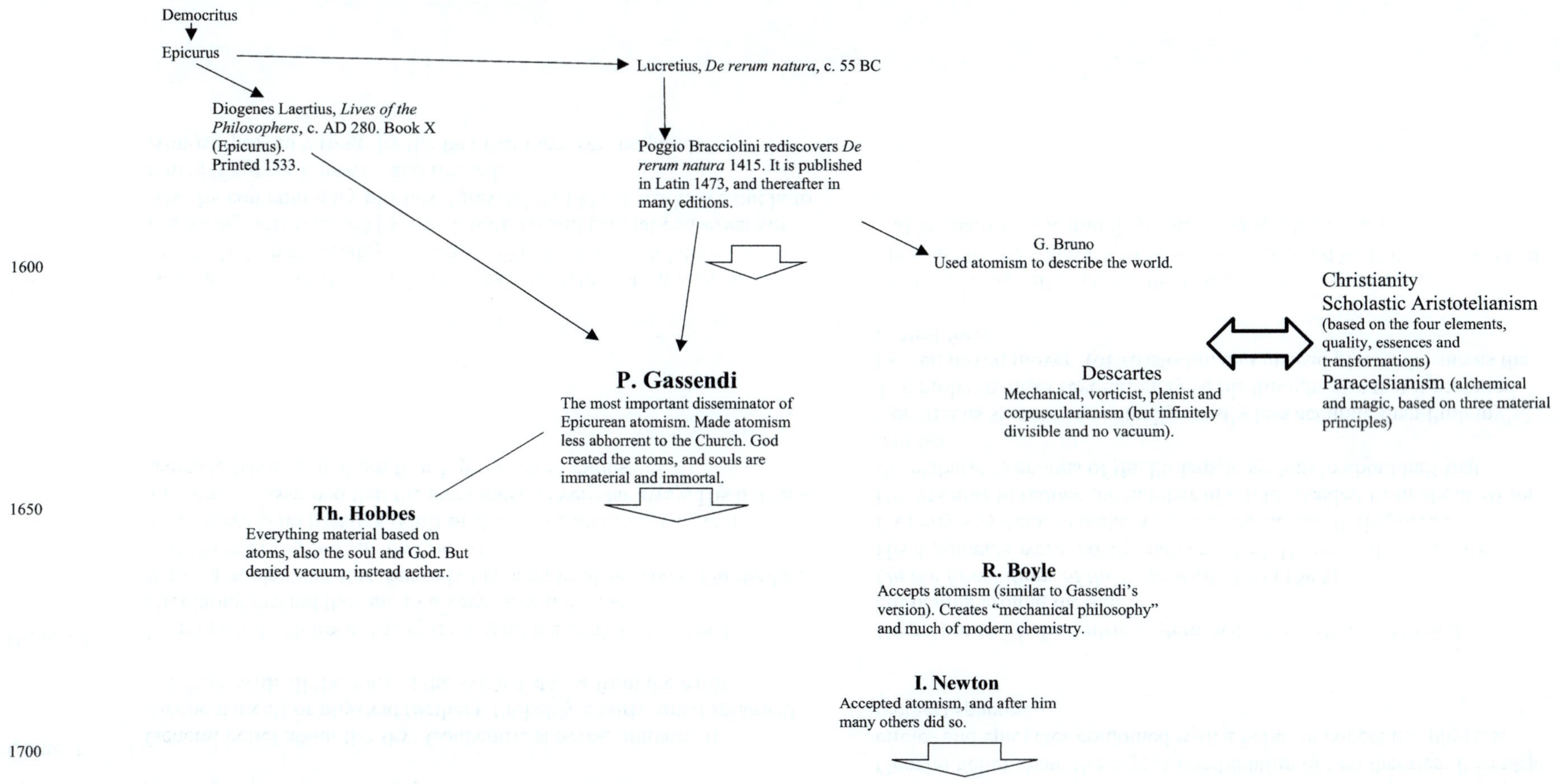

The Revival of Epicurean Atomism during the Seventeenth Century: 2. More Names

At the beginning of the century, two different forms of element teachings, the four-element scholastic Aristotelianism and the three-principle system of Paracelsus, ruled in regard to the nature of matter. By the latter parts of the century, atomism had taken over the place as the most acceptable theory of matter.

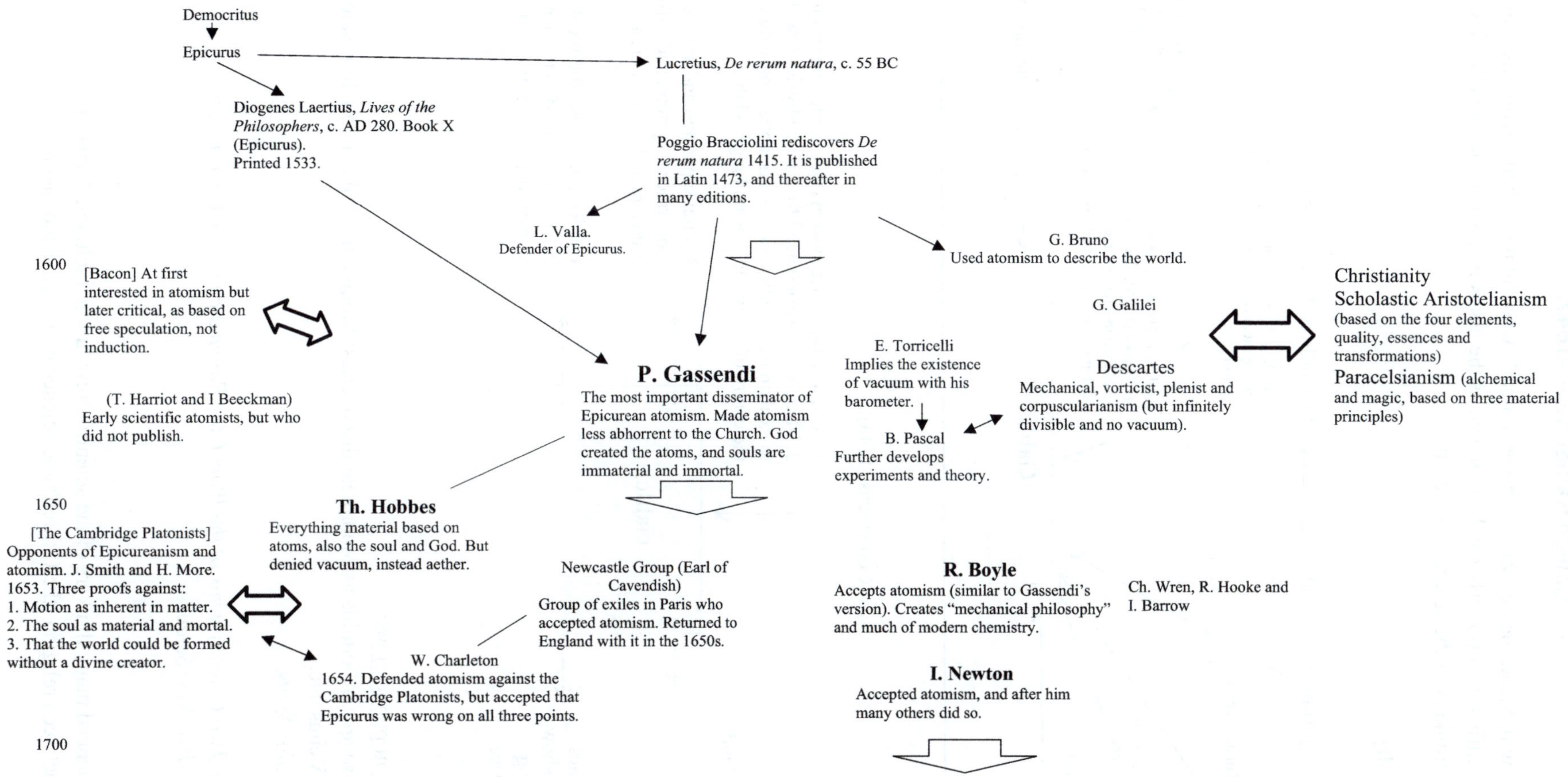

Galileo Galilei (1564–1642)

Galilei, contemporary with Shakespeare, was a "universal man", who did more than anyone else for the ushering in of the scientific revolution. He can be regarded as the last Renaissance man and the first or most important of the early thinkers of the scientific revolution.

Major Influences on Galilei:

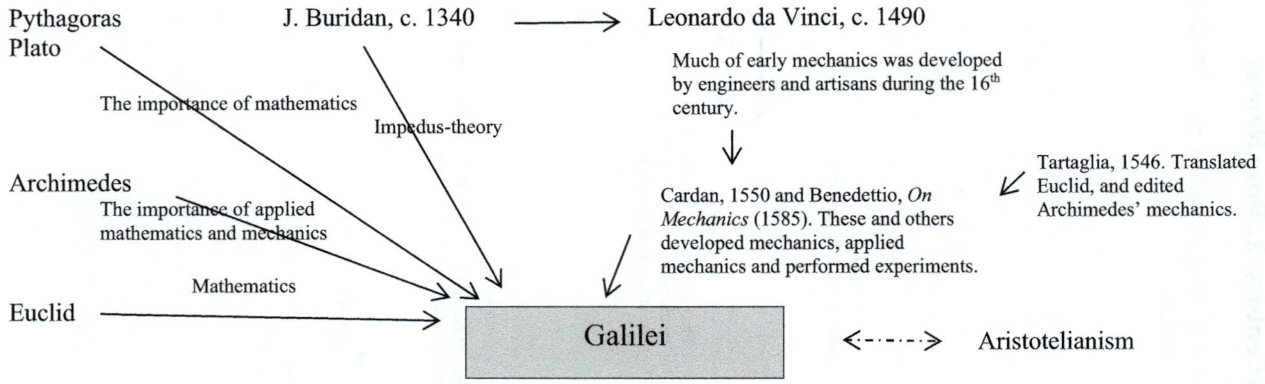

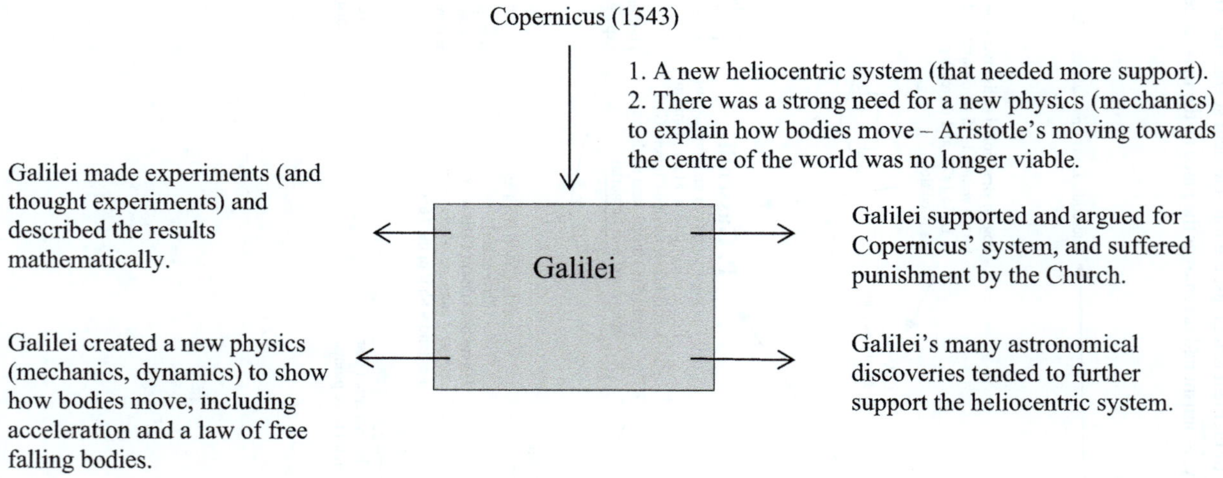

Galilei's most important publications:

1610: *The Starry Messenger*. About his astronomical discoveries: Mountains on the moon, four satellites of Jupiter, the phases of Venus, etc.

1612: *Letters on the Solar Spots.*

1623: *The Assayer*

1632: *Dialogue on the Two Chief Systems of the World, the Ptolemaic and the Copernican*

1638: *Discourse upon Two New Sciences*

Consequences:

1. Galilei greatly influenced many thinkers of the scientific revolution, including Newton.
2. Made the conflict between religion (the Church) and science more acute and visible.

Isaac Newton (1642–1727)

1642. Newton born in Woolsthorpe, Linconshire, England.

Newton graduated 1665.
Summer 1665 - spring 67. Cambridge University closed due to the plague.

1666. Newton invents the differential and integral calculus (which elements he called fluxions and inverse fluxions), suitable for calculating limits, dealing with infinitesimals and for adding infinite series – extremely useful for mathematics applied to physics. Leibniz independently created a similar calculus shortly afterwards, but published it before Newton, which led to an unpleasant battle for who discovered it first.
Developed modern optics – the composition of light (spectrum) and dispersion.
Invented a new type of telescope – the mirror telescope (1668).

1669-95. Newton was professor of mathematics at Cambridge.

1672. Published short text on light and colour.
Received criticism and promises never to publish again.

Thereafter, for 10-20 years:
Gave lectures and worked on:
- alchemy
- biblical prophecy
- biblical chronology

1679. Based on an idea of R. Hooke (circular movement can be understood as a combination of tangential motion and a falling towards the centre) and the laws of Kepler, Newton realized that the force on planets in elliptical orbits must vary as the inverse square of the distance from the sun – but did not publish (or inform anyone).

1687. *Philosophiae naturalis principia mathematica* (*Mathematical Principles of Natural Philosophy*) with its new concept of mass, the principle of inertia, and his three laws of motion. From these he deduced the orbits of the planets and other things, including the shape of the earth and the explanation of the tides. With this work Newton thus presented his theory of gravity and joined terrestrial and stellar mechanics.

c. 1695. Moved to London and became master of the mint.

1704. *Opticks: or A Treatise of the Reflection, Refraction, Inflections and Colours of Light*.
This work summarizes his optical discoveries and his corpuscular theory of light. In the influential "Queries" at the end, a long series of hints and suggestions regarding physical and chemical phenomena, often combined with suggested experiments which set the stage for much of scientific thought and experimentation during the 18th century. Among much else, he here suggested the highly influential hypothesis that forces (gravity, electrical, heat, light) are mediated by weightless aethers or imponderable fluids.

Newton revised and expanded *Opticks* in three further editions.

1727. Newton dies 85 years old.

Science and Scientific Thought during the Enlightenment (c. 1700–1790) and the Romantic Era (c. 1790–1830)

After the dramatic progress in science and scientific thought during the seventeenth century, the Enlightenment, the eighteenth century, can seem as something of a disappointment, with much less dramatic development. However, much of the Enlightenment, and the whole Enlightenment project, was a direct consequence and application of the developments during the Scientific Revolution. Several of the Enlightenment thinkers spoke of wanting to apply the geometric spirit (the scientific spirit) to the whole of society, and most of the heroes of the Enlightenment were the main representatives of the Scientific Revolution.

Map over Eighteenth Century Natural Science: Main Schools and Traditions

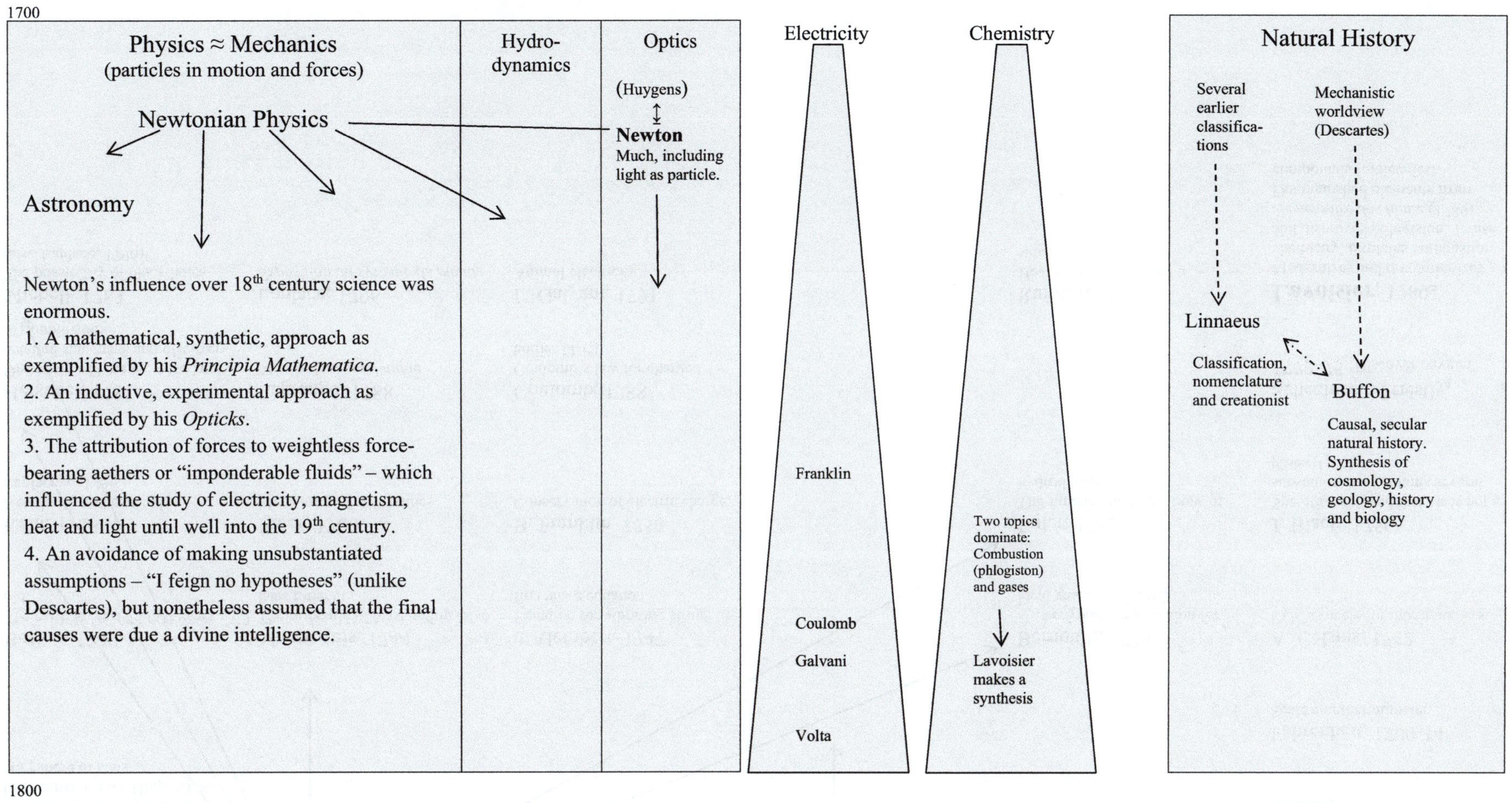

Table over the Most Important 18th Century Natural Philosophers (Physicists and Chemists)

Astronomy and Cosmology	Mechanics	Electrostatics	Optics	Hydrodynamics	Chemistry
	[Newton]		[Huygens, 1690 Wave-theory of light]		[R. Boyle]
Halley, 1705 *A Synopsis of the Astronomy of Comets*. Predicted the return of comet (which vindicated Newton) and showed that stars have moved since Hipparchos. No sphere of stars.			Newton, *Opticks* 1704. Particle theory of light.		Stahl, 1697, 1711 Introduces phlogiston as explanation for combustion. Will color the whole century of chemistry.
					Fahrenheit, 1709-14 Scale and thermometer.
Buffon, 1749 The earth at least 75 000 years old.	Maupertuis, 1744 The principle of least action (also later Euler).	d'Alembert, 1747 Equation for vibrating string, the first wave equation.		Bernoulli, 1738 *Hydrodynamica*, equation for ideal gases and liquids.	A. Celsius, 1742 Celsius-scale for thermometers
Kant, 1755 Nebular hypothesis. Also Laplace, c. 1796.	Euler, 1760 Mechanics of stiff bodies	B. Franklin, 1750 Conservation of electric charge.		Euler, 1755 The fundamental equations of hydrodynamics.	J. Black, 1756 Specific heat, and that air is not a substance, but contains several gases (CO_2)
Herschel, 1781 Discovery of Uranus. Also later detailed star-maps and discovery of double-stars.	Lagrange, 1788 *Méchanique analytique*	Coulomb, 1788 Coulomb's law for charged bodies ($1/r^2$).			Scheele and Priestly, 1772-75. Discover oxygen
Michell, 1783 The possibility of black holes (also Laplace, 1796)	Laplace, 1796 *Exposition du système du monde*	L. Galvani, 1791 Animal electricity.		Rumford, 1797 Heat is movement, not a substance.	**Lavoisier**, 1780s. Modernizes and revolutionizes chemistry. Explains combustion and disproves phlogiston. *Traité élémentaire de chimie* (1789). Distinguished elements from compounds (molecules).

Map over Romantic Natural Science (*Naturphilosophie*) and Its Prehistory

The main tradition in the history of modern science is the mechanistic, materialistic and mathematical one, which dominated especially in England and France. A (mostly) minor tradition, which existed alongside it, was one which emphasized vitalism, and was more qualitative, anthropocentric and idealistic, and which blossomed especially in Germany during the Romantic era. It is often referred to as *Naturphilosophie*.

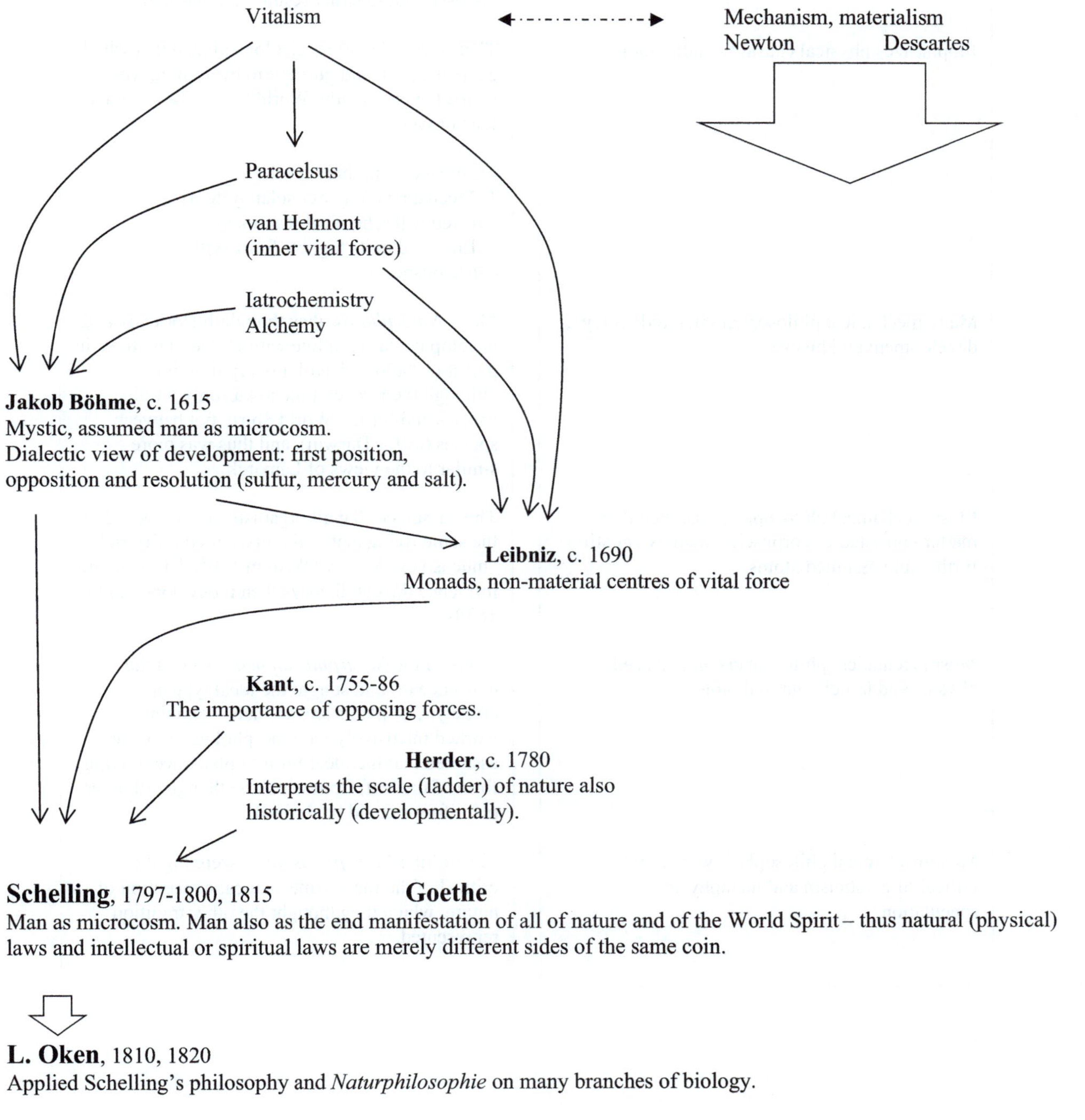

The Worldview of *Naturphilosophie* contra that of Mechanistic Natural Philosophy

Mechanical Natural Philosophy	*Naturphilosophie*
Emphasizes matter, mechanical principles (direct contact) and especially motion.	Strong emphasis on forces rather than matter, and that those forces are "polar", that they often are opposed, but together resulting in harmony.
Emphasizes physical causality and motion.	Their view was often teleological, and that which gave direction and purpose to everything was called the eternal, the World Spirit, the Abstract Idea or God.
	Nature exists in three levels: 1. Mechanical (e.g. our solar system) 2. Chemical (chemical reactions) 3. Life (and in its highest form, with consciousness).
Many mechanical philosophers ignored biology, development and history.	Many *Naturphilosophen* thought historically and developmentally, in line with Herder, and thus, in part foreshadowed evolutionary thinking – although their development was only within species and for the World Spirit, not between species (as for Darwin), and thus was more similar to the views of Lamarck.
Most mechanical philosophers assumed that matter consisted of corpuscles, many went still further and assumed atoms.	They assumed that all organism are composed by the same living material units, called infuserial "mucus vesicles" by Oken, in 1805. This came to influence the cell theory when it developed in the 1830s.
Most mechanical philosophers emphasized physics and largely ignored biology.	Many of the *Naturphilosophen* assumed the existence of, and sought the ideal type or archetype of plants and animals – and thus worked intensively with morphology. Goethe assumed that the ideal form of plants were in the form of the leaf, from which he thought all other plant forms could be derived.
Most mechanical philosophers were highly critical of mysticism and metaphysical speculation.	Most of the *Naturphilosophen* were highly critical of the merely mechanical conception of nature, which they thought the other tradition represented.

Map over Romantic Natural Science (*Naturphilosophie*), c. 1790–1830

(Part Natural Science, Emphasizing Immaterial Polar Forces, and Part Philosophy)

Much of romantic science and *Naturphilosophie* began in Germany, and especially based on the philosophy of Friedrich Schelling and his *Ideas for a Philosophy of Nature* (1797) and Hegel's *Encyclopedia of the Philosophic Sciences* (1817), but later spread outside Germany.

Naturphilosophie and romantic science were critical of the reductionist, mechanistic and naturalistic science of the Scientific Revolution and of the Enlightenment:

Neoplatonism and Christianity
Paracelsus and Kant
Schelling and Hegel

Bacon and Newton	⟺	*Naturphilosophie* was instead speculative, idealistic and holistic
Inductive and empiristic methods. Atomism and mechanism. The importance of quantification and mathematics.	⟺	Influenced by Neoplatonism and Kant (the mind shapes nature or how we see nature). Emphasized immaterial forces rather than matter (atomism) – everything being manifestations of a single underlying force – often teleological. Nature consists of opposites or polarities, but which fundamentally are united.

Romantic scientists expressed little interest in traditional physics (mechanics), but were fascinated by electricity and magnetism; the imponderables, especially heat and light (later to become thermodynamics and electrodynamics); and anatomy and morphology

J.W. Ritter, *Physics as Art* (1806) Hans Christian Ørsted. Discovered electromagnetism in 1820. ↓ Ampère Humphry Davy Michael Faraday	Goethe, *Zur Farbenlehre* (1810), in direct opposition to Newton.	An emphasis on the relatedness in anatomy and nature, combined with idealism and teleology. Goethe, *On the Metamorphosis of Plants* (1790) ↕ Lorenz Oken, *Elements of Physio-Philosophy* (1805-11) Richard Owen.

Others were interested in psychology and medicine, as, for example, Carl Gustav Carus, romantic painter and physician, who wrote about the soul and the unconscious, as well as his follower, von Schubert, *Symbolism of Dreams* (1814).

Science and Scientific Thought during the Modern Era, c. 1800–2020

Overview of Modern Physics 1800 – 2020: 1. Characteristics

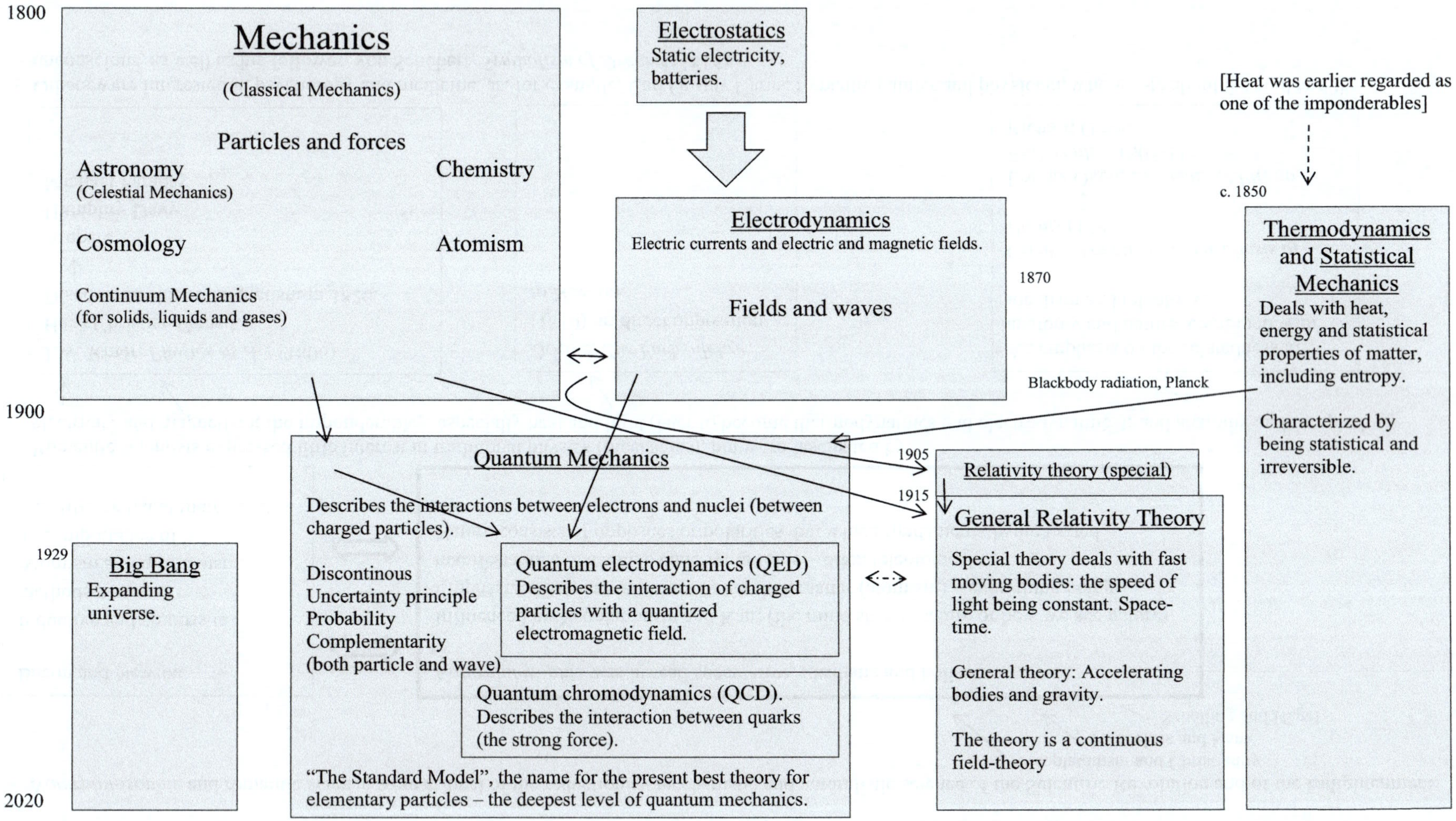

Overview of Modern Physics 1800 – 2020: 2. Main Representatives

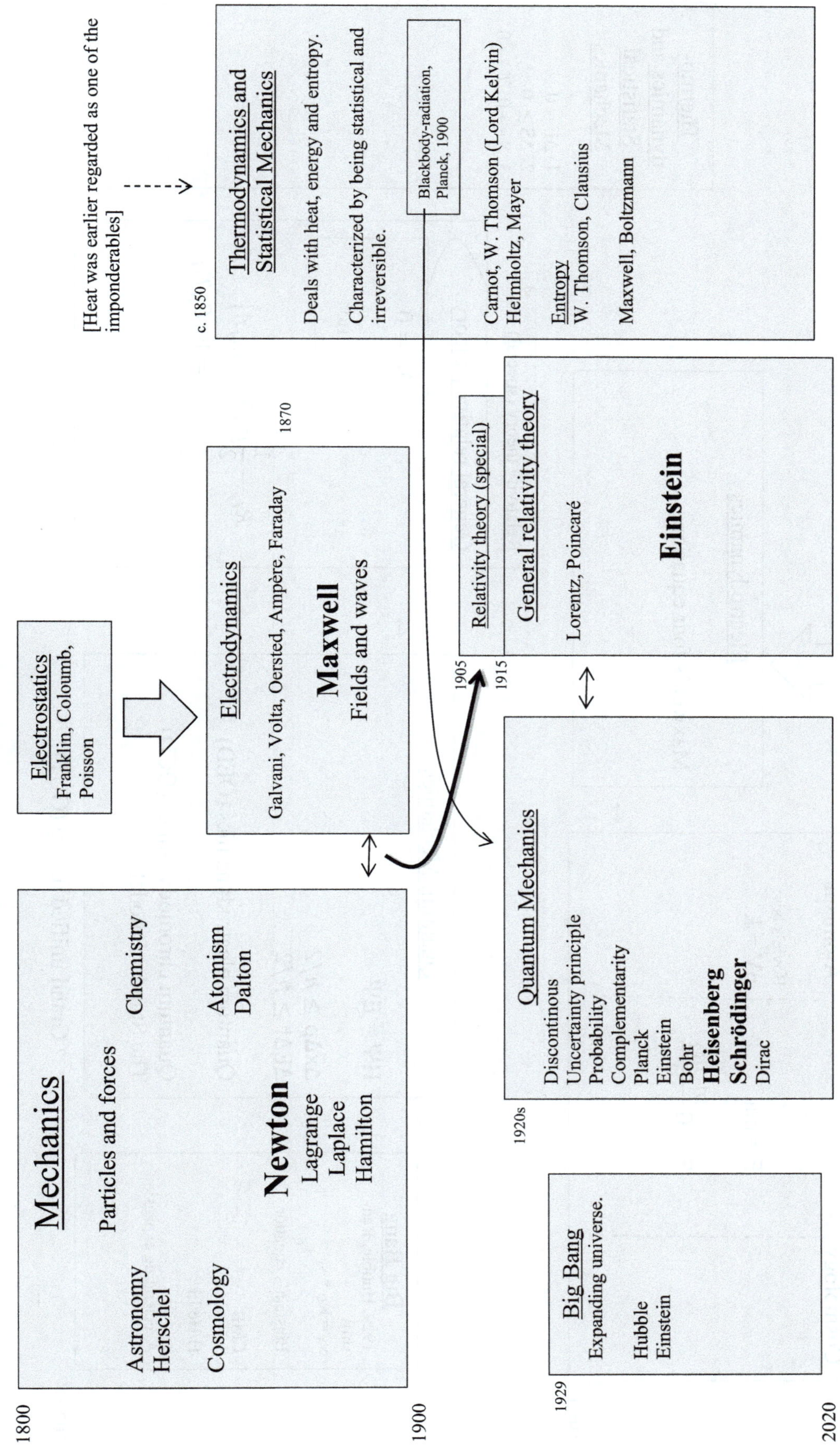

Overview of Modern Physics 1800 – 2020: 3. Scientific Formulas

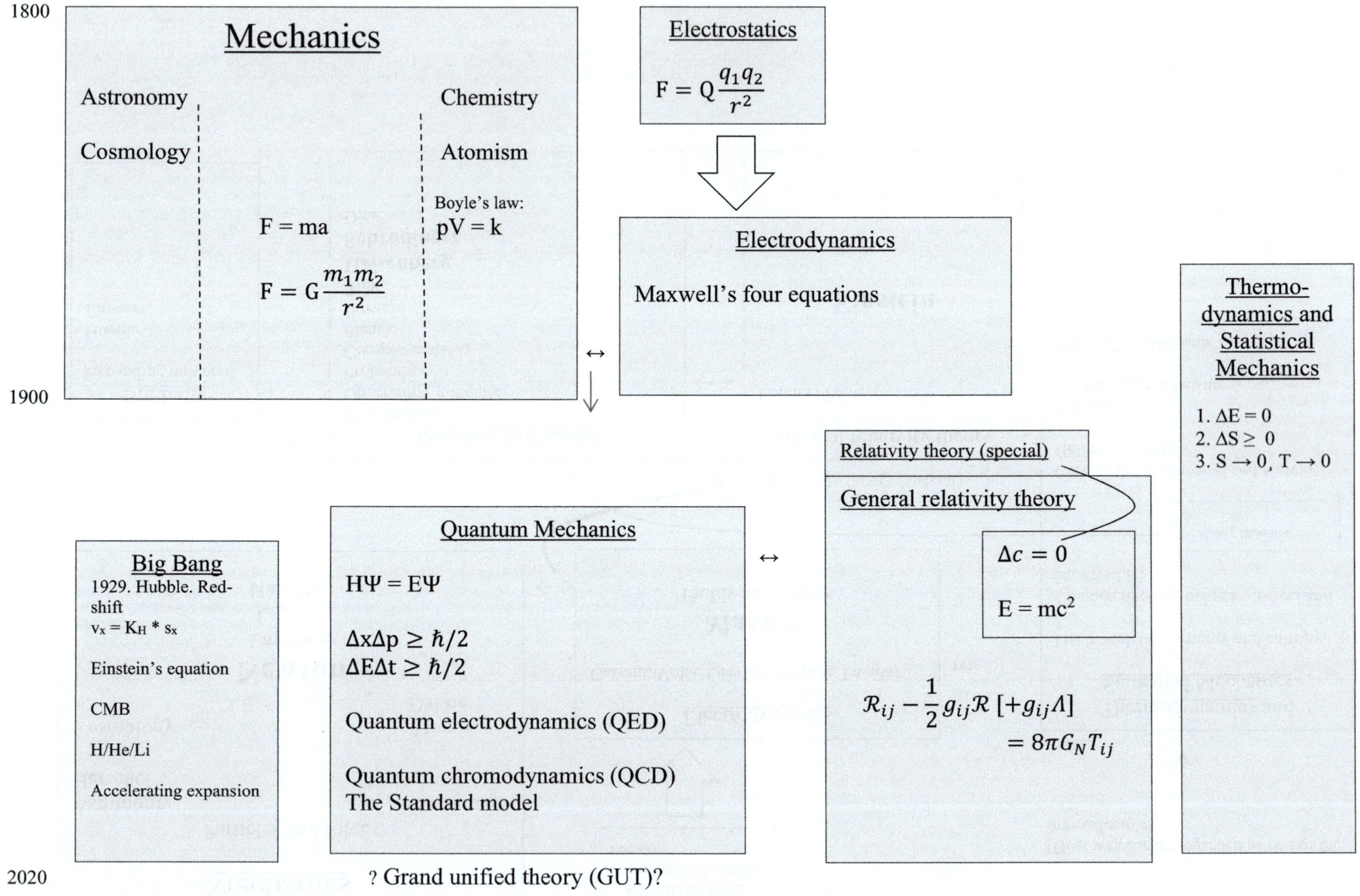

Overview of Modern Physics 1800 – 2020: 4. Worldviews and Conceptual Structures

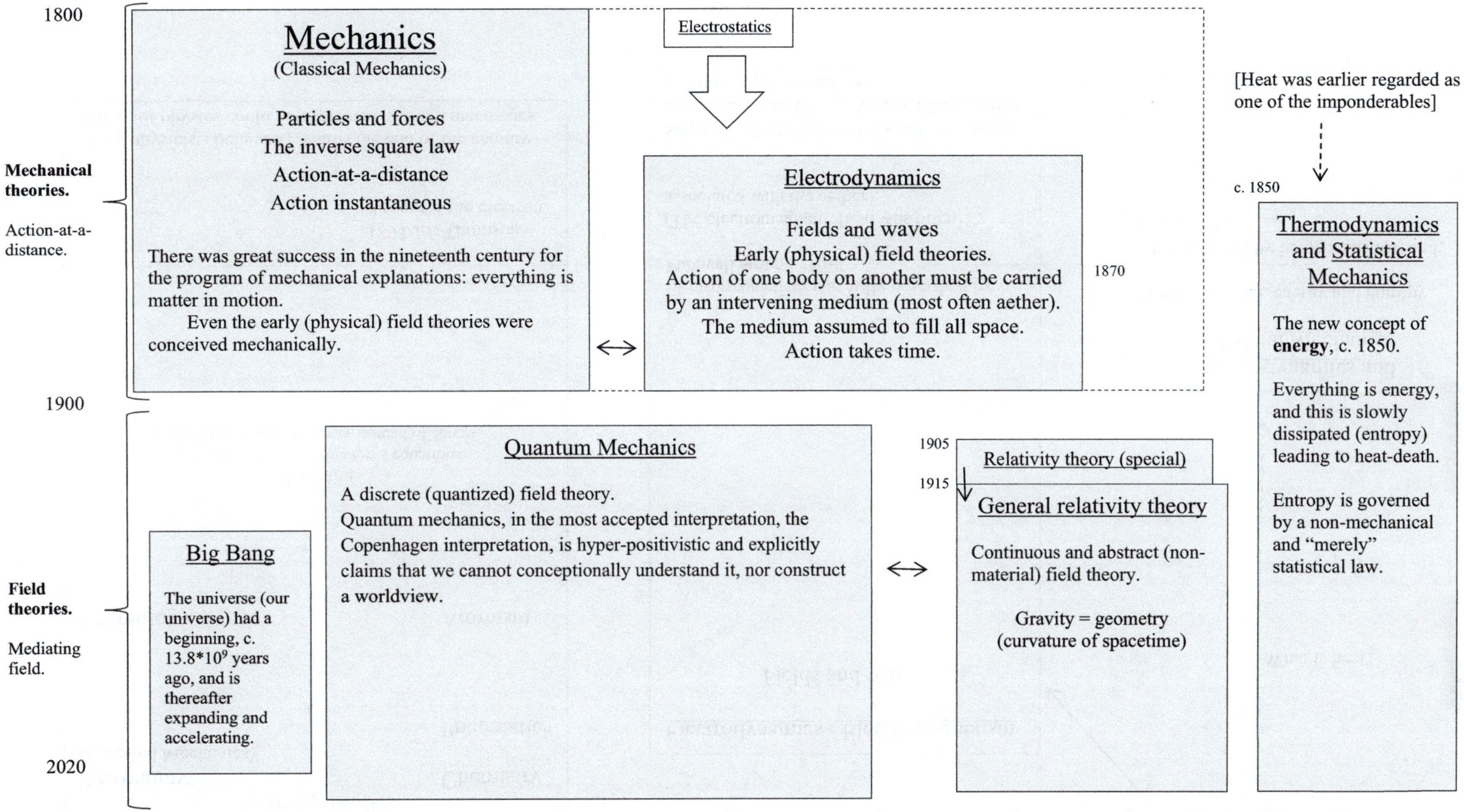

Overview of Nineteenth Century Physics: 1. Characteristics

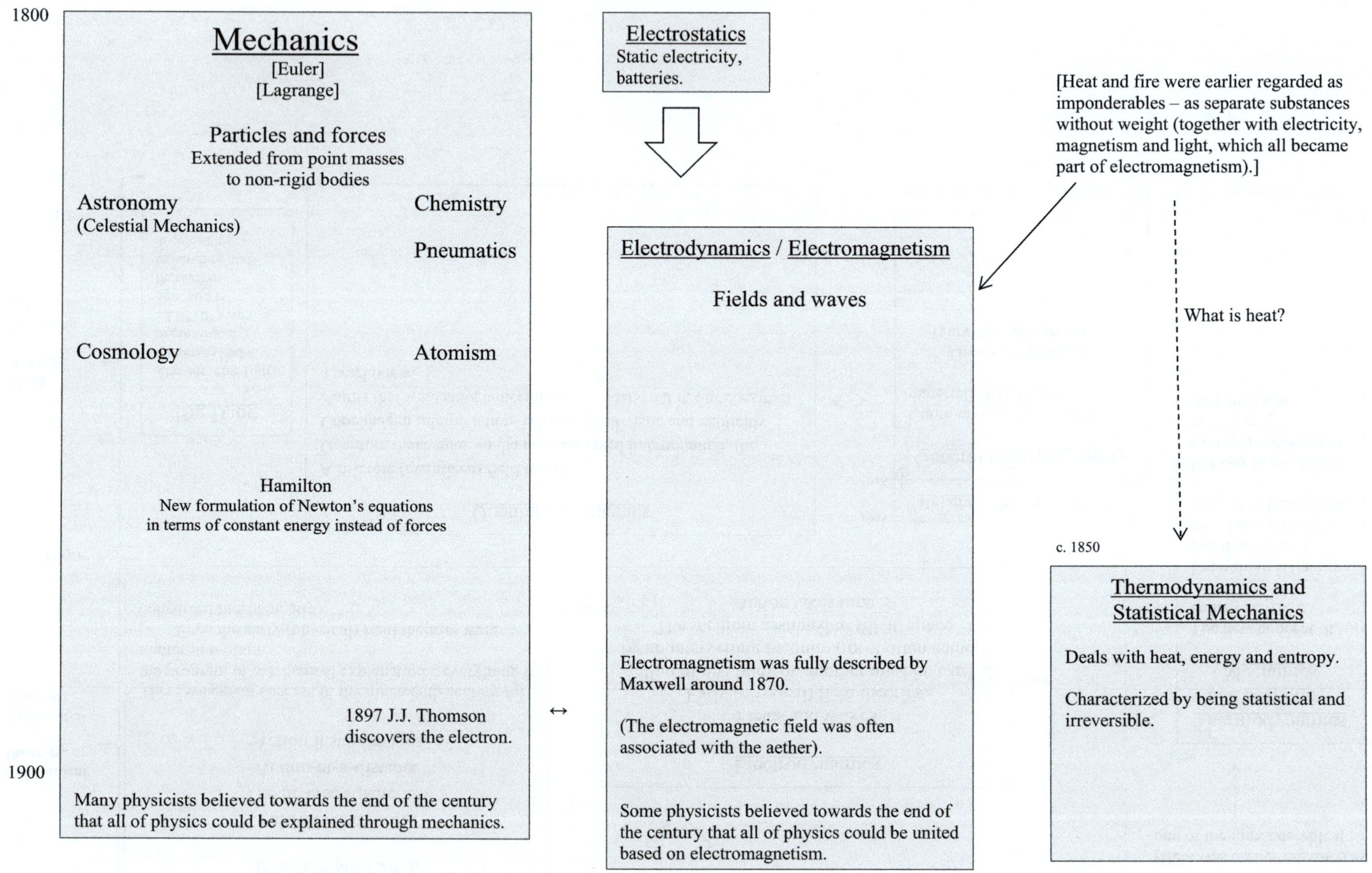

Field Theories in Nineteenth Century Physics

The concept of field theory emerged in the nineteenth century – initially as a mechanical agency to transport energy over distances, to avoid the problem with action-at-a-distance – especially applied for the non-central forces discovered in electricity and magnetism.

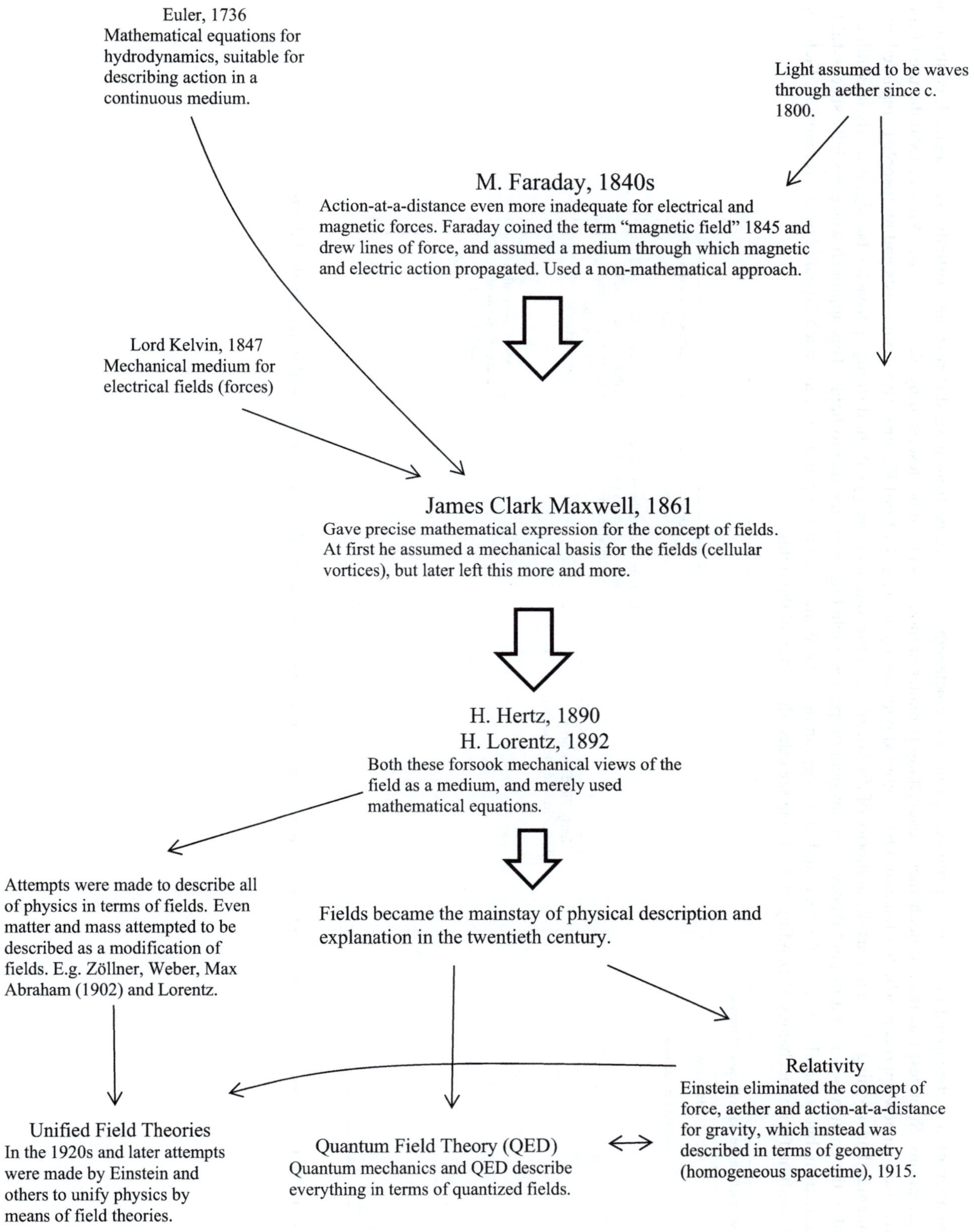

Attempts at Conceptual Unification of Physics during the Nineteenth Century: 1. Outline

The attempt at a conceptual unification of physics during the nineteenth century came from three different and partially, as it seemed, contradictory fundamental assumptions. Strongest was that of mechanics – everything is matter in motion and can be described by Newton's laws. After c. 1850 the use of the concept energy gave rise to a new approach, which often could be combined with mechanics (e.g. by Lord Kelvin). These physicists and philosophers regarded energy as more fundamental than anything else (and thus the authority of the law of the conservation of energy). A third strong approach – but significantly less influential until the very end of the century – was that of field theory – all physical properties, including mass, are due to fields. This approach was especially successful for electricity and magnetism, but could also be used for heat (and in the beginning of the century similar approaches were much favoured by the romantics). Field theories could also be combined with mechanics, depending on how one regarded the nature of the fields.

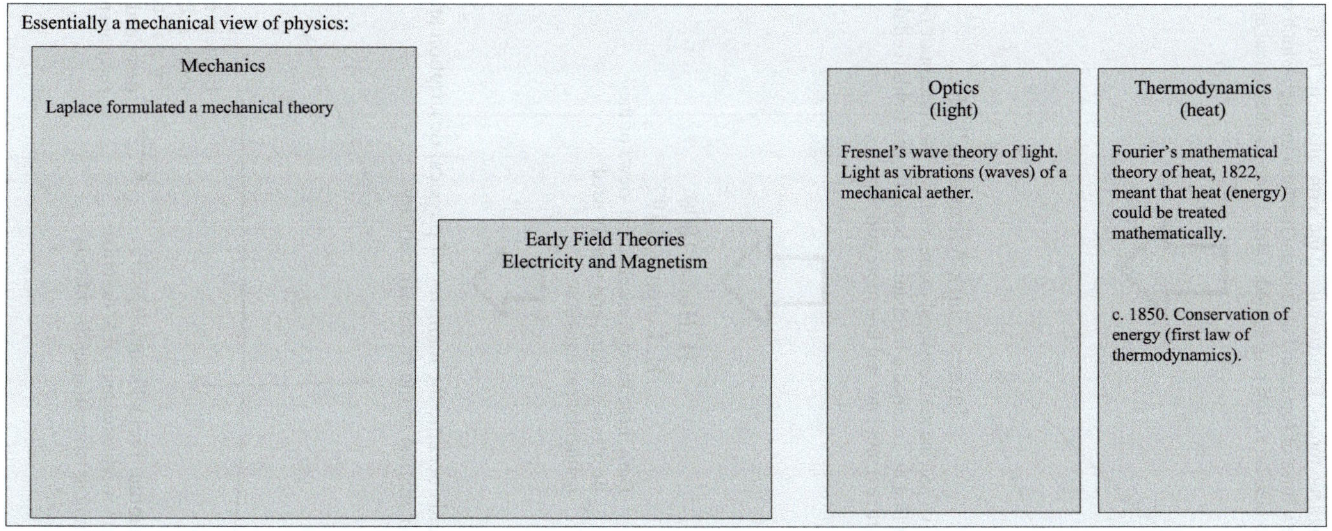

Post 1900. Field theories take over, and now with a much more abstract sense of field (no longer physical or mechanical fields).

Table over the Most Important 19th Century Physicists and Chemists: 1. Names

Astronomy and Cosmology	Mechanics	Electrostatics and Electrodynamics	Optics	Thermodynamics	The Structure of Matter (Atomism)	Chemistry
	Lagrange, 1788 Laplace, 1796	Volta, 1800	Young, 1802		**Dalton**, 1803	Proust contra Bertollet
		Oersted, 1811	Fresnel, 1812		Avogadro, 1811	H. Davy, 1810
		Ampère, 1820		Carnot, 1824	Brown, 1827	Berzelius, 1820
1838. Parallax measured.	Hamilton, 1834	Faraday, 1830			Faraday, 1835	Wöhler, 1828
1846. Neptune discovered.		Joule, 1840		1840-1850. **First law of thermo-dynamics**: Discovered by several persons: Carnot, Joule, Mayer, Helmholtz		Justus Liebig, 1825-1850
Kirchhoff and Foucauld, 1859			Foucauld, 1850		Kirchhoff and Bunsen, 1859	Frankland, 1852
				Clausius, 1865		Kekulé, 1858
						Cannizzaro, 1858
		Maxwell, 1868		Maxwell-Boltzmann's law, 1871		**Mendelejev**, 1869
Michelson-Morley, 1887.		Herz, 1887		**Boltzmann**, 1877	Michelson-Morley, 1887.	
					Röntgen, 1895	
					Becquerel, 1896	
					J.J. Thomson, 1897	

Table over the Most Important 19th Century Physicists and Chemists: 2. Summaries

Astronomy and Cosmology	Mechanics	Electrostatics and Electrodynamics	Optics	Thermodynamics	The Structure of Matter (Atomism)	Chemistry
[Newton]	[Newton]	[Gilbert, 1600]	[Newton's atomic theory of light dominated the 18th century]			
[Kant-Laplace] [Herschel]	[Lagrange 1770-1800] [Laplace, 1770-1800]	[Franklin, 1750] [Coulomb, 1788]		[Rumford, 1797]		[Lavoisier, 1780s]
1800.................	....	....	....	....	....	....
						Chemical reactions: The nature of chemical compounds (molecules). Proust (fixed proportions) contra Bertollet (infinite series of proportions).
	Lagrange, 1788 Laplace, 1796 Both continue to develop aspects of Newtonian physics.	Volta, 1800 First batteries	Young, 1802 Discovers interference, and explains it by proposing light as waves.		**Dalton**, 1803 Atomic theory (not fully accepted until the end of the century). Assumed definite proportions. *New System of Chemical Philosophy* (1808). Calculated relative atomic weights.	
		Oersted, 1811 Discovered the magnetic field created by an electric current.	Fresnel, 1812 Explains diffraction		Avogadro, 1811 At the same V, T, P different gases contain the same number of particles (atoms). Also by Ampère in 1814. Not accepted by most chemists for some generations	H. Davy, 1810 Electrochemistry etc.
		Ampère, 1820 The relation between electricity and magnetism. Sometimes regarded as the founder of electrodynamics		Carnot, 1824 Heat and work are interchangeable	Brown, 1827 Brownian motion	Berzelius, 1820 Detailed work on elements and atomic weights in support of Dalton and the atomic theory + chemical nomenclature. After electrochemical work he assumed forces between atoms and molecules to be electrostatic.
1838. Distance to stars measured by parallax.		Faraday, 1830 Discovers induction			Faraday, 1835 He and many others began work on Faraday flasks	Wöhler, 1828 Synthesized organic substance (urea) from inorganic. No life force.
1846. Neptune discovered.	Hamilton, 1834 Better mathematical formulation of classical mechanics	Joule, 1840 Examines the relation between electrical current, resistance and heat		1840-1850. The **first law of thermodynamics**: Discovered by several		Justus Liebig, 1825-1850. Influential German organic chemist.

persons: Carnot, Joule, Mayer, Helmholtz

Foucauld, 1850
Shows that speed of light greater in air than water. Supports wave-theory of light

Frankland, 1852
Valency

Kirchhoff and Foucauld, 1859
Explain how spectra from stars and the sun work. This led to intensive work on astronomical spectra.

Clausius, 1865
Expresses first two laws of thermodynamics

Kirchhoff and Bunsen, 1859
Began using optical spectroscopy for chemical analysis in the laboratory.

Kekulé, 1858
(and others). Introduced the chemical bond and chemical structure.

Cannizzaro, 1858
Rediscovered Avogadro's work, distinguished atoms from molecules and realized that absolute atomic weights could be measured.

1870-1900. Intensive spectroscopic studies of many stars – results possibly suggesting that stars evolve and transmutation of elements – though this conclusion rarely accepted.

Maxwell, 1868
Synthesis of electrodynamics in four equations

Maxwell-Boltzmann's law about the distribution of molecular speeds in gases, 1871

Mendelejev, 1869
The periodic table. Predicted three new elements and their properties. (Several others had similar ideas.)

Michelson-Morley, 1887. No evidence for ether.

1895. Helium found on earth (earlier from spectra from the sun)

Herz, 1887
Discovers electromagnetic waves

Boltzmann, 1877
Statistical formulation of entropy

Michelson-Morley, 1887. No evidence for ether.

Röntgen, 1895
X-rays

Becquerel, 1896
Radioactivity

J.J. Thomson, 1897
Discovers the electron, and in 1899 its mass and charge

Table over the History of Chemistry and the Chemical Bond 1800–1950: 1. Summaries

Organic	Inorganic	Electrochemistry	Atomic theory	Chemistry of gases
	Wenzel, Richter, c. 1780	[Coulomb, 1788]		[Lavoisier, 1780s]
1800.................	....	....	....	....
Bertollet ⟷	Proust			
Chemical reactions: The nature of chemical compounds (molecules). Proust (fixed proportions) contra Bertollet (infinite series of proportions).				
		Volta, 1800 First batteries	**Dalton**, 1803 Atomic theory. Assumed definite proportions. *New System of Chemical Philosophy* (1808). Calculated relative atomic weights.	
				Gay-Lussac, 1809 Gases combine in simple volume proportions
		H. Davy, 1810 Electrochemistry etc.		Avogadro, 1811 At the same V, T, P different gases contain the same number of particles (atoms). Not accepted by most chemists for some generations.
		Berzelius, 1820 Detailed work on elements and atomic weights in support of Dalton and the atomic theory + chemical nomenclature. He assumed forces between atoms and molecules to be electrostatic.		
Wöhler, 1828 Synthesized organic substance (urea) from inorganic. No life force.				
Justus Liebig, 1825-1850. Influential (German) organic chemist.				
	Frankland, 1852 Valency			

The Chemical Bond

During the 18th century not enough detailed chemical knowledge was known to understand bonding. Many chemists used Newton's concept of attractive short-range "affinities" or the closely related "chemical affinities" to explain chemical reaction in general.

Early in the 19th century both the atomic theory of Dalton and electricity were examined, and several chemists, including Berzelius, assumed that chemical bonding was due to electrostatic attractions, c. 1810-1890.

However, the view had difficulties, especially to explain how many elements could bond to themselves, e.g. $O + O \rightarrow O_2$, as well as for many organic molecules.

A new different and more structural approach was began by Frankland in 1852 and developed especially by Kekulé, for both inorganic and organic molecules. This was based on that each element seemed to combine in specific proportions, e.g. H, 1; O, 2; N, 3 and C, 4. This was first called "affinity units", then valency, and then chemical bond – and Kekulé and others used it successfully to construct a theory of chemical structure (e.g. benzene).

However, there were no successful way to explain how chemical bonding actually worked – both gravity and electrostatics seemed excluded.

Kekulé, 1858
(and others). Introduced the chemical bond and chemical structure.

Kirchhoff and Bunsen, 1859
Began using optical spectroscopy for chemical analysis in the laboratory.

Cannizzaro, 1858
Rediscovered Avogadro's work, distinguished atoms from molecules and realized that absolute atomic weights could be measured.

Mendelejev, 1869
The periodic table. Predicted three new elements and their properties. (Several other chemists had similar ideas.)

1874. Structural chemistry: Pasteur, T'Hoff, Achille Le Bel.

J.J. Thomson, 1897
Discovers the electron, and in 1899 its mass and charge

↓

c. 1930. Quantum chemistry developed.

With the discovery of the electron, 1897, new approaches were made, including by J.J. Thomson, letting electrons circulate in shells – to correspond to the 8-pattern of the periodic table. This was further developed by W. Kossel, G.N. Lewis and I. Langmuir.

Langmuir distinguished two major types of bonds, ionic and covalent – in the former case an electron is donated and the bond was electrostatic, in the latter case the atoms or molecules shared the electrons. How this worked was not understood until after quantum mechanics was developed, 1925-27.

Close to 1930, Linus Pauling and others developed valence-bond theory, based on quantum mechanics. Mullikan developed an alternative quantum mechanical approach based on molecular orbitals, and after the second world war it gradually displaced Pauling's model.

Map of the History of Chemistry 1750–2000: Overview

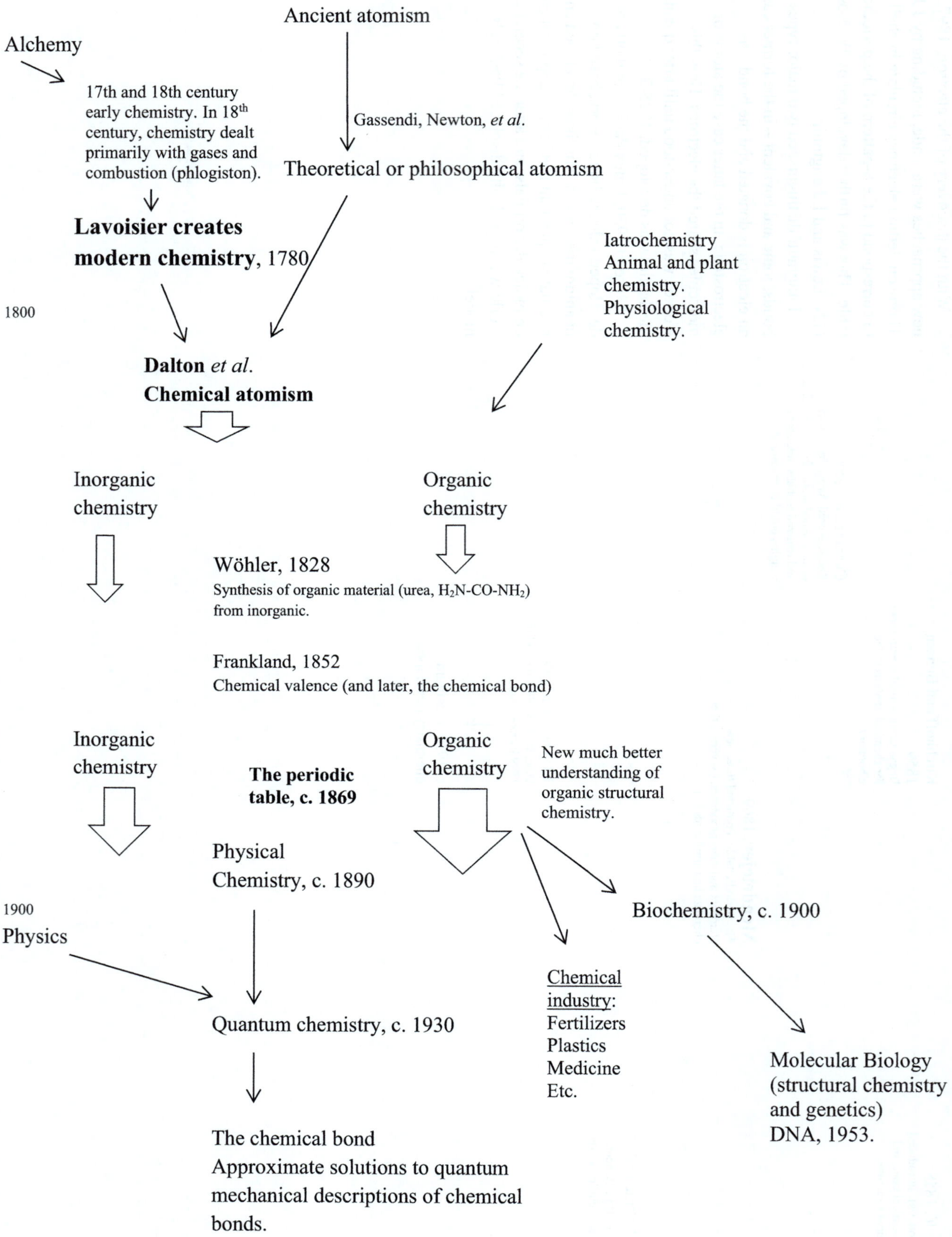

Modern Physics, 1900–2020

Overview of Twentieth Century Physics: 1. Characteristics of the Main Fundamental Theories

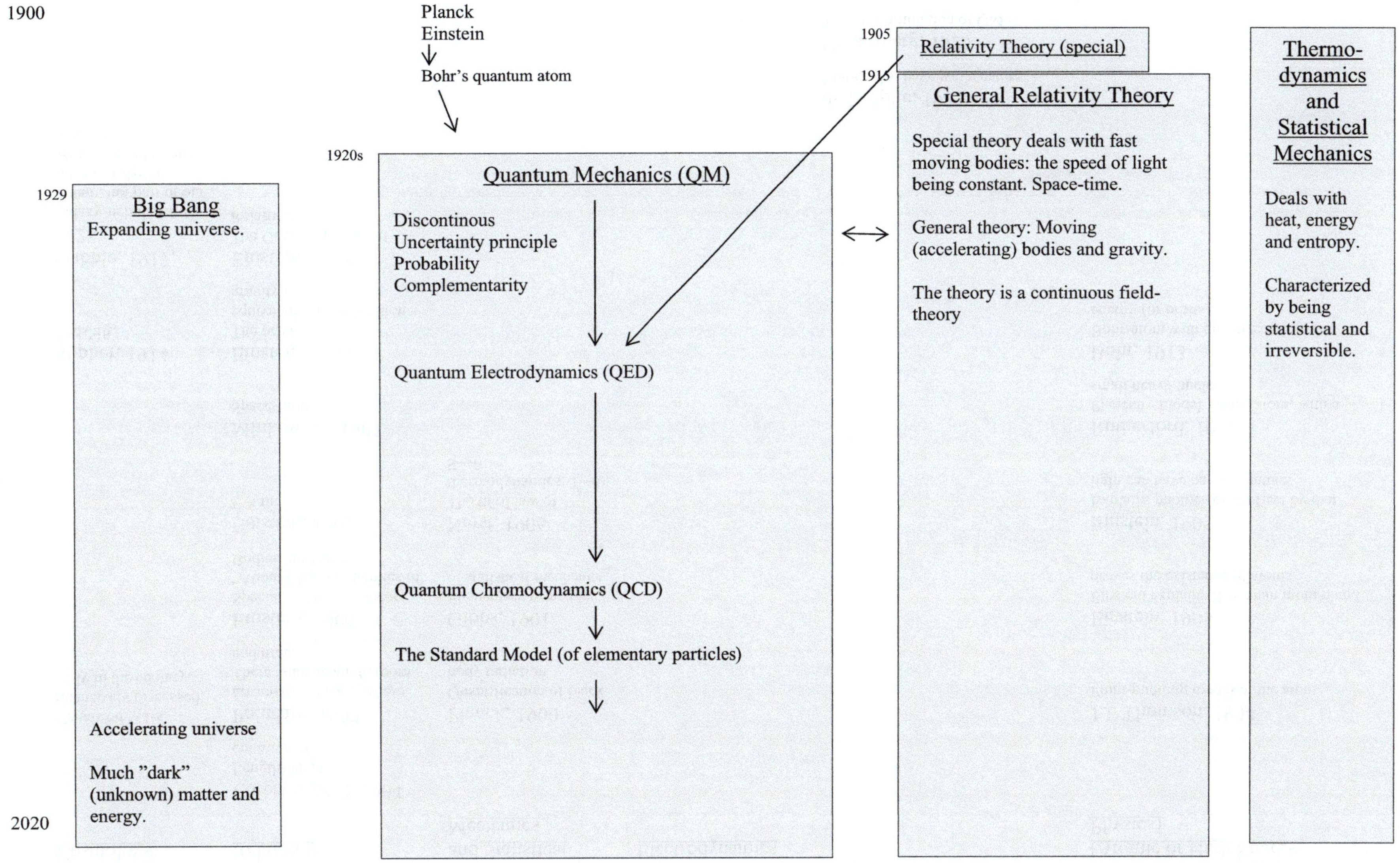

Table of the Most Important 20th Century Physicists: 1. Overview

Astronomy and Cosmology	Mechanics → Relativity	Thermodynamics and Statistical Mechanics	Electrostatics and Electrodynamics	Quantum Mechanics	The Structure of Matter (Atomic or High Energy Physics)
	Lorentz, 1892-1904 Length- and time contraction.				
Canals on Mars intensively discussed early in the century.	Poincaré, 1902 *La science et l'hypothèse*. There is no absolute room and time.	Planck, 1900 Quantification of black body radiation.			J.J. Thomson, 1904 Plum-pudding model of the atom.
	Einstein, 1905 Special Relativity Theory. "About Electrodynamics of Bodies in Motion"	Gibbs, 1901 Elementary principles of statistical mechanics.			Einstein, 1905 Einstein explains Brownian motion and proves the existence of atoms.
	Einstein, 1905 $E = mc^2$	Nerst, 1906 The third law of thermodynamics: $T \to 0$, $S \to 0$			Einstein, 1905 Explains photoelectric effect by that light can have particle nature.
	Minkowski, 1907 Space-time.				Rutherford, 1911 Planetary model for the atom, with a small heavy nucleus.
Slipher, 1914 Redshift.	Einstein, 1911 The principle of equivalence (acceleration = gravity)				Bohr, 1913 Bohr-atom with quantized electron energy (or orbits).
Hubble, 1917, 1928 Galaxy-nebulae are distant, not part of our galaxy. Relation between redshift and distance.	Einstein, 1915 The General Theory of Relativity.				
				de Broglie, 1923 Matter also have wave nature.	
				Heisenberg, 1925 Matrix-formulation of QM.	
				Schrödinger, 1926 Differential equation formulation of QM. The wave-function.	

Born, 1926
Probabilistic interpretation of the wave-function.

Heisenberg, 1927
The Uncertainty principle.

Dirac, 1928-31
The Dirac equation. Includes special relativity. Predicts the positron.

Chadwick, 1932
Discovers the neutron.

Anderson, 1932
Discovers the positron.

Fermi, 1934
Theory of β-radioactivity. Introduces weak interaction.

Yukama, 1935
Introduces the strong interaction.

1945. Atomic bomb.

Feynman, 1948
Feynman-diagrams and renormalization.

Higgs, 1964
Higgs introduces the Higgs field – particle found in 2012.

Gell-Man, 1964
Introduces quarks

1969. First human landing on the moon.

History of Relativity Theory

Albert Einstein (1879–1955): 1. Outline

Special theory of relativity, 1905
In June 1905 Einstein published the article "On the Electrodynamics of Moving Bodies", 31 pages (pp), which constitutes his special theory of relativity.
It is based on two axioms: the speed of light in vacuum is constant and physical laws must be independent of which system of reference is chosen. The second assumption is the same as the assumption of Newton. Space and time becomes space-time. Dismissed aether.

General theory of relativity, 1915
Gravity = acceleration.
Gravity = curvature of space-time (geometrized gravity)

1917. General theory of relativity applied to cosmology, "Kosmologische Betrachtungen zur allgemeinen Relativitätstheorie", the beginning of modern theoretical cosmology.

Einstein's work on light (and other work) was one of the most important contributions to the development of quantum mechanics.

<u>Einstein was also active in a number of fields outside physics</u>:

He was a dedicated pacifist. Nonetheless, he wrote an important letter to the American president and asked him to develop the atomic bomb, so as to do this before the Germans.

He was a sionist (and was offered to become the first president of Israel, but declined).

He was an atheist (at least did not believe in a personal God), but sometimes used religious language.

Albert Einstein (1879–1955): 2. More Details

Einstein's doctoral dissertation was published 30 April 1905: *A New Determination of Molecular Dimensions* (Univ. of Zürich). In 1905 he also published four seminal papers in the journal *Annalen der Physik*:

1. "On a Heuristic Point of View Concerning the Production and Transformation of Light", 17 March 1905, 17 pp.

Light can also have particle nature, later called photon. It is then easy to explain and understand the photo-electric effect.

An important ingredient of quantum mechanics.

Nobel Prize 1921

However, it was not until after the Compton effect was discovered in 1923, and well explained by Einstein's work, that people began to accept that light could sometimes behave as a particle. The "light particle" was named photon in 1926.

2."On the Movement of Small Particles Suspended in Stationary Liquids Required by the Molecular-Kinetic Theory of Heat", May 1905, 10 pp.

Continuation of his Ph.D., and one of the strongest proofs that liquids consist of atoms (or molecules).

3. "On the Electrodynamics of Moving Bodies", June 1905, 31 pp.

Special relativity theory. Space and time becomes space-time. The speed of light as constant. Dismissed aether.

General theory of relativity, 1915
Gravity = acceleration
Gravity = curvature of spacetime
(geometrized gravity)

1917. General theory of relativity applied to cosmology. The result is unstable (universe as expanding or contracting), so Einstein adds a cosmological constant to force it to become stable and static: "My greatest blunder". This was the beginning of modern theoretical cosmology.

4. "Does the Inertia of a Body Depend upon its Energy Content?", Sept. 1905, 3 pp.

This is the first, but different, formulation of the expression: $\mathbf{E = mc^2}$

Einstein spent much of the last 30 years of his life in attempts to unify relativity and quantum mechanics (from a relativity perspective) without succeeding. These two main theories continued and continue to be fundamentally different and separate.

General theory of relativity ⟷ **Quantum mechanics**

He also spent much time criticizing the standard (Copenhagen) interpretation of quantum mechanics – "God does not play dice" – especially its probabilistic character. Much of this critique has been important to further develop and in interpreting quantum mechanics, but the Copenhagen interpretation remains the standard interpretation.

Hidden variable and other interpretations. Einstein, Schrödinger, Bohm	⟷	Copenhagen interpretation of quantum mechanics. Bohr, Heisenberg

Relativity Theory 1900–1920

At the end of nineteenth century, the two main theories about nature were not compatible.

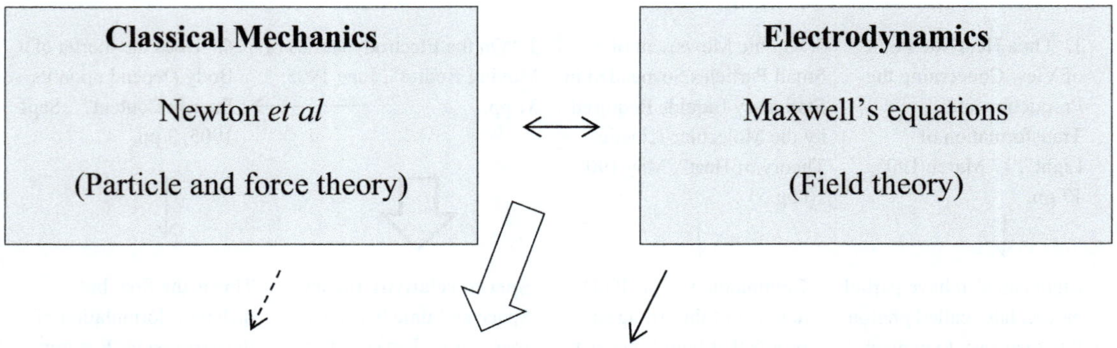

In part, as a response to this, Einstein in 1905 makes some basic assumptions, which led to the special (limited) relativity theory (sRT):

1. Invariance – the same laws must be valid in all reference systems.
2. The velocity of light in vacuum is constant ($\Delta c = 0$) for all observers, from any reference system.

↓ Consequences (after Einstein's thought-experiments)

Absolute space and time → spacetime
Absolute simultaneity → makes no sense any longer (but within a reference system simultaneity is still valid)
Space, time and mass are affected by velocity
The twin paradox

The special relativity theory is limited (to objects travelling in straight lines with constant speed) and Einstein worked hard to expand it to the general relativity theory (GRT), 1908-1915, which is mathematically much more complex than the special relativity theory.

Assumptions:
1. Gravity and acceleration cannot be distinguished
2. Non-Euclidean geometry

↓ Consequences (after Einstein's thought-experiments)

Gravity as action-at-a-distance disappears and is explained by non-Euclidean curvature of spacetime.

Only relatively few physicists, including Planck and Eddington, realized quickly that this was new and valuable. For the public, relativity theory was proved right with the solar exclipse 1919 and the result that light is bent by the sun (as predicted by Einstein).

In 1917 Einstein applies relativity theory to the universe, which is the beginning of modern (theoretical and mathematical) cosmology.

The History and Development of Relativity Theory (Theory of Gravity)

Relativity theory was developed by Einstein, and, unlike quantum mechanics, remains essentially unchanged since then (1915/16).

Einstein (and others) did not realize the physical possibility of black holes (but the concept of black holes followed from his equations). That is an area in which much new work has been done since the death of Einstein.

The cosmic microwave background radiation (CMB) has given us new information about the universe (it is essentially "flat", i.e. Euclidean) and some of Einstein's discussions and solutions concerning the development of the universe, although not incorrect, are no longer relevant.

Since the discovery of dark energy (the increasing acceleration of the universe), c. 2000, this may indicate the need to extend Einstein's theory (which so far can be done by re-introducing the "cosmological constant").

Einstein's theories predicted gravitational waves, ripples in the curvature of spacetime, unlike the case for Newtonian physics. Gravitational waves were discovered for the first time in 2015, and announced in 2016 (and received the Nobel prize in 2017).

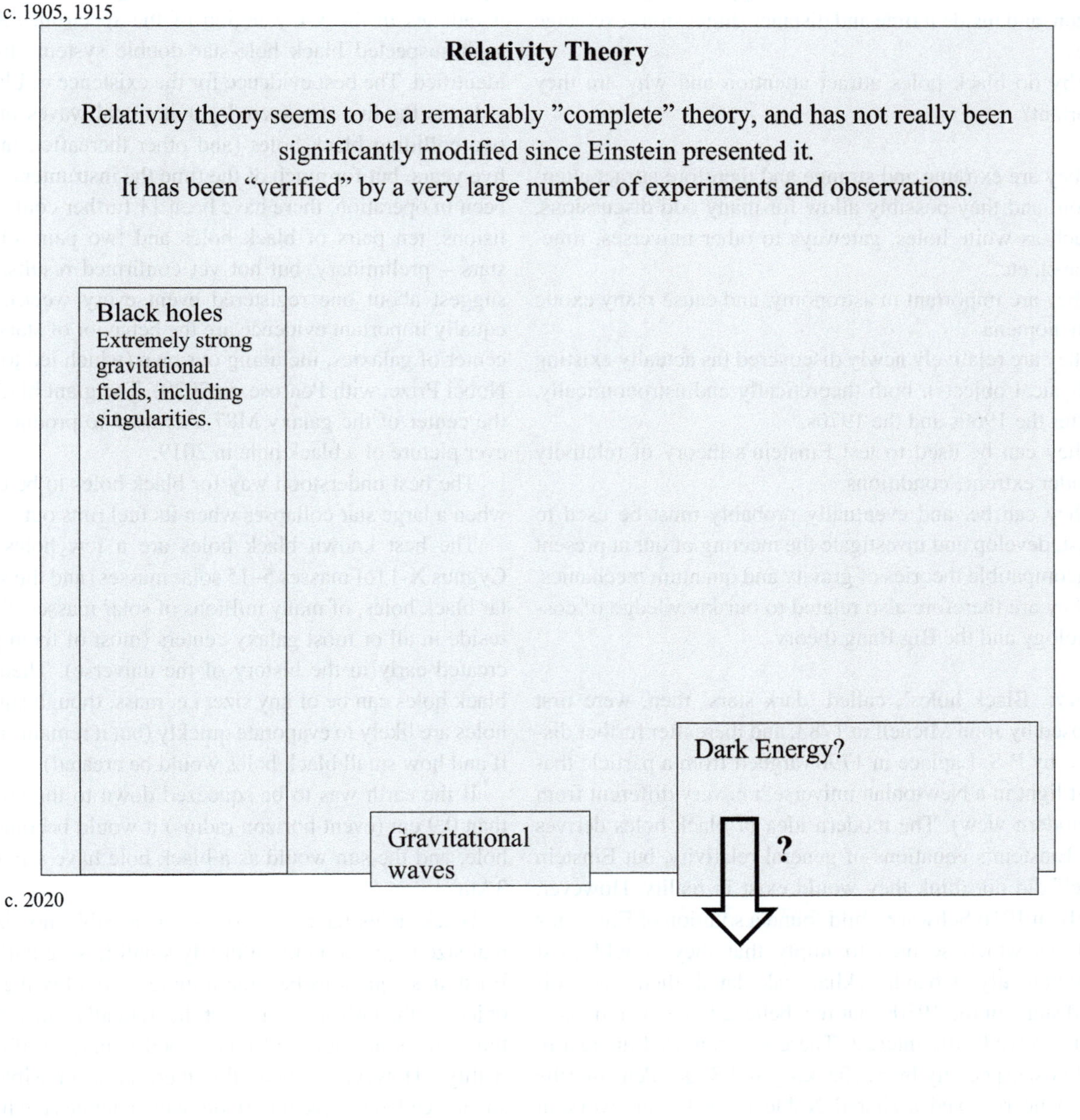

What are Black Holes and Why Are They Important?

Black holes are regions of spacetime where gravity is so strong that nothing can escape, not even light (the curvature of spacetime is infinite). This leads to a number of, for us, paradoxical consequences. They cannot be directly seen, and yet, the brightest known objects in the universe are due to them; supernovas (big stars which explode, some of them then become black holes), much more luminous, quasars (an active supermassive black hole, usually at the center of a galaxy) and blazers (quasar, with jets out from the poles directed directly at us). They are perhaps the most massive things in the known universe, and at the same time also perhaps the smallest things. Furthermore, for an external observer, time slows down and stands still when approaching the event horizon, and inside it time and distance dimensions exchange places.

Why do black holes attract attention and why are they important?

1. They are extreme and strange and therefore attract attention, and they possibly allow for many odd discussions, such as white holes, gateways to other universes, time-travel, etc.
2. They are important in astronomy, and cause many exotic phenomena.
3. They are relatively newly discovered (as actually existing physical objects), both theoretically and astronomically, after the 1960s and the 1970s.
4. They can be used to test Einstein's theory of relativity under extreme conditions.
5. They can be, and eventually probably must be used to test, develop and investigate the meeting of our at present incompatible theories of gravity and quantum mechanics. They are therefore also related to our knowledge of cosmology and the Big Bang theory.

History: 'Black holes', called 'dark stars' then, were first proposed by John Michell in 1783, and thereafter further discussed by P-S. Laplace in 1796 (argued from a particle theory of light in a Newtonian universe, i.e. very different from the modern view). The modern idea of black holes derives from Einstein's equations of general relativity, but Einstein himself did not think they would exist in reality. However, already in 1916 Schwarzschild found a solution of Einstein's equations which seemed to imply that they should exist mathematically. Chandrasekhar calculated them, as collapsed stars, in the 1930s, but few believed they would exist, and it received little interest. There was renewed interest in the 1960s, especially by S. Hawking and Roger Penrose (the latter, who received a shared Noble Prize for his work in 1964 in 2020). John Wheeler, or a student of his, is often credited as having coined the term 'black hole' in 1967, but there seems to have been several sources for this at that time. Since then they have become a major topic of study in physics, astronomy and astrophysics.

Black holes cannot radiate or reflect light, so they are not directly visible. However, their existence can be established by other, mostly indirect means, such as gravity lensing, their influence on nearby bodies (usually a companion star), gas being heated before being sucked into it or by gravitational waves.

The first object assigned as a probable black hole was Cygnus X-1 in 1972 (and it is still thought to be a black hole, it radiates in the x-ray region of the spectrum). About 50 such suspected black hole-star double systems have been identified. The best evidence for the existence of black holes today is the first discovered gravitational waves in 2015 of two colliding black holes (and other thereafter, in the first five years, but for much of the time the instruments have not been in operation, there have been 11 further confirmed collisions, ten pairs of black holes and two pairs of neutron stars – preliminary, but not yet confirmed results, seem to suggest about one registered event every week). Another equally important evidence are the behavior of stars near the center of galaxies, including our own (which led to a shared Nobel Prize, with Penrose, in 2020). The giant black hole at the center of the galaxy M87 was used to produce the first ever picture of a black hole in 2019.

The best understood way for black holes to be created is when a large star collapses when its fuel runs out.

The best known black holes are a few holes like the Cygnus X-1 (of masses 5–15 solar masses) and the superstellar black holes, of many millions of solar masses, thought to reside in all or most galaxy centers (most of them probably created early in the history of the universe). Theoretically, black holes can be of any size, i.e. mass, though small black holes are likely to evaporate quickly (but it remains uncertain if and how small black holes would be created).

If the earth was to be squeezed down to the size of less than 0.9 cm (event horizon radius) it would become a black hole, and the sun would as a black hole have a radius of c. 3 km.

Black holes have 'two sizes' or possibly no size. Their real size is just a point, infinitely small (a singularity, when Einstein's equations become infinite – possibly the smallest object in the universe – at least theoretically, since the quantum effects are not well understood it may be different in reality). However, outside this there is an invisible border, sphere, called the event horizon, which defines the limit from which no light can exit or return, which usually is what is referred to as the size of a black hole (although actually all matter, mass, is concentrated to the infinitely small point at the centre) and referred to as the Schwarzschild radius. Considering the complexity of the mathematics of general relativity and of black holes, the expression for the Schwarzschild radius is remarkably simple: $R = 2GM/c^2$ (where G is the gravitational constant, M the mass of the black hole and c the speed of light).

Outside it there is another special distance, 1.5 times the Schwarzschild radius, where light can circulate the black hole in a stable orbit (just as a planet does its sun).

Black holes in some ways are simple. They can only have 3–4 properties; mass, charge, angular momentum (rotation), and probably, if Hawking was right, also temperature (due to quantum tunneling, but extremely low temperatures for larger black holes, only a very tiny fraction of a Kelvin). In this brief account I only discuss non-charged and non-rotating black holes.

What makes black holes complicated to understand, is the complex mathematics of general relativity and of quantum mechanics, which here also needs to be combined. The complication includes both the solving of the equations, or simplified versions of them, and the interpretation what this actually means.

What makes black holes especially interesting to theoretical physicists is that they constitute the best known place, together with the Big Bang, where Einstein's general theory of relativity meets quantum mechanics. These two theories are both very successful and accurate, but cover different areas of space and mass, and are not compatible, do not meet well. Usually they do not meet at all since gravity is so weak that it can be ignored on atomic scales, while quantum mechanics can be ignored for normal and astronomical scales – except for in the Big Bang and black holes.

Black holes seen from the outside:

We now (2023) have knowledge of about ten to fifty stellar (collapsed stars) black holes, ten pairs of collided black holes (known due to gravitational waves) and several super-stellar black holes (but assumed to reside at the centre of all or almost all galaxies). These numbers are likely to increase rapidly. Theoretically, a black hole is infinitely small, but the region from which light cannot escape (the event horizon) depends on its mass, and is usually referred to as its size.

For an outside observer viewing an object moving towards a black hole, time will go slower and slower (since time is affected by gravity), and at the event horizon stand still, and the object can therefore not be seen reaching further (but due to redshift, dopplershift, it will have become invisible to us before then anyway). However, the object itself would not notice any border or horizon, but simply continue to fall towards the central singularity (but after it has passed the event horizon it is impossible ever to escape the black hole).

Black holes seen from the inside:

Black holes are even more fascinating and mysterious when seen from the inside (but this seems impossible to be empirically confirmed since nothing is able to leave the inside of a black hole). Our knowledge of the laws of physics break down. There are two 'insides', inside the event horizon, and 'inside' the central singularity. When an object comes inside the event horizon, the spacetime changes, so that the three space dimensions become time dimension(s), while the singe time dimension becomes a space dimension. What happens "inside" the central singularity is not known. Many assume that it is not a true singularity, since quantum effects needs to be taken into account.

A major topic of discussion and research today is 'the black hole information paradox', which concerns that information seems to be lost in black holes, but this should not be allowed according to quantum mechanics. Information here meaning quantum numbers which should be conserved. There seems at present to exist no genuine consensus of how this paradox is to be resolved.

A Black Hole

A black hole really has no size at all. It seems to be infinitely small (symbolized by the point). However, it has an invisible horizon, the event horizon, and anything entering that cannot escape again. Nonetheless, the black hole is visible by the fact that it affects the spacetime and any matter nearby.

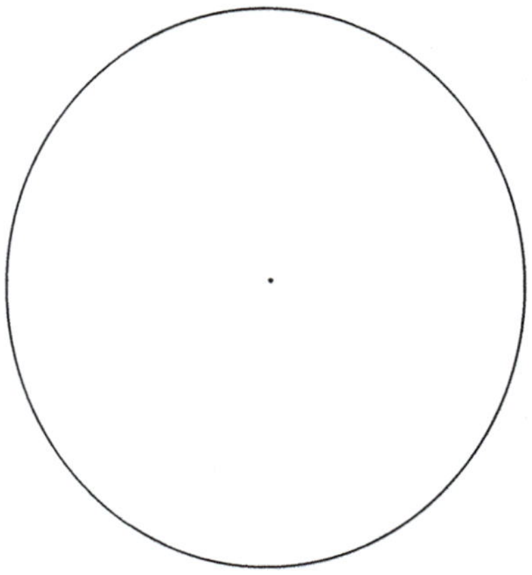

The Relations Between Classical Mechanics, Quantum Mechanics and Relativity Theory

Both relativity theory and quantum mechanics include classical mechanics in the sense that they give the same results when describing large systems moving slowly (much below the speed of light). However, for such systems, classical mechanics are still used, for it is much simpler mathematically. It is only under rather special conditions (for us) that it is necessary to use the two more modern theories, which are not compatible with one another. This is one of the major problems in modern physics.

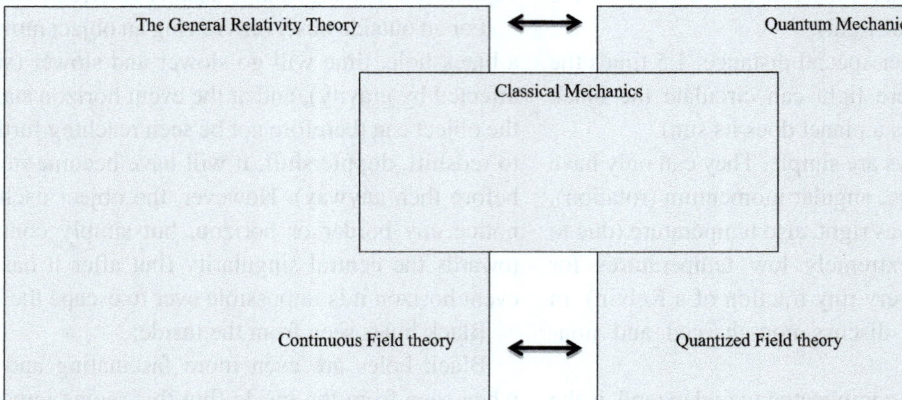

Relativity theory and quantum mechanics are not compatible and not in agreement with one another. In practice this is rarely a problem, since gravity plays a negligible role in quantum reality (the sub-microscopic world), and the quantum effect is negligible in the macroscopic world, but when working on the Big Bang Theory or black holes, and conceptually, this is a major disadvantage.

The Relation Between Classical Physics and Relativity Theory

Classical Physics (1600–1900)	**Relativity Theory (Post-1915)**
Valid for	Valid for:
Used to be believed to be valid for everything and everywhere. Nowadays, regarded as a good approximation for medium sized objects moving relatively slowly.	Larger than microscopic objects. Slow and fast moving objects, also near $c = 3*10^8$ m/s.
Assumes nature to be continuous.	Continuous field (consistent with nature as continuous).
Absolute space and time, but all motion is relative.	There is no absolute space or time, but spacetime is absolute.
Space (3-D) and time are independent of each other.	Spacetime (space and time are variable and dependent of one another).
The world is three-dimensional.	The world is four-dimensional.
The dimensions are basically fixed and constitute a stable stage upon which everything happens.	The four-dimensional "stage" is continually dynamic and changes with every change in the world (on the stage).
Things can occur simultaneously here and on the other side of the universe. Since time is absolute, there is no problem with the simultaneous.	Things can occur simultaneously if both events are in the same spacetime (are near to one another), but at enormous distances or when at very different speeds, simultaneity cannot be ensured.
All speeds are possible.	The speed of light in vacuum is the maximum speed.
There is no obvious relation between energy and mass, but conservation of mass and energy separately.	$E = mc^2$. Matter and energy can be transformed into each other, but their sum is conserved.
Deterministic.	Deterministic. Non-probabilistic. Theoretically exact.
Relatively easy to understand (but action-at-a-distance has no explanation).	Difficult to understand, but not impossible (different from our everyday world). Action-at-a-distance solved by the presence of a spacetime field.
Determined with mathematics and empiricism.	Determined by thought-experiments, mathematics and empiricism.

The Best Known and the Most Important Physical Equation: 1. Introduction

$$E = m \bullet c^2$$

The equation shows that matter (mass) and energy can be transformed into each other, and that their total sum is conserved. It also says that since c^2 (c is the speed of light) is an enormously large number, the energy content of matter is very large.

One of the surprising things about it is that it is so simple.

A more complete version is the following:

$$E = \sqrt{\left(m_0^2 c^4 + p^2 c^2\right)}$$

The last term, pc, is needed for particles like light (photons) which have a zero restmass. p stands for momentum (mass times velocity).

$$m = m_0 / \sqrt{\left(1 - v^2 / c^2\right)}$$

where m_0 is the rest mass.

This equation was discovered by Einstein. He hinted at it in his publication about the special relativity theory 1905, and shortly thereafter, in September 1905, he publishes a short three page paper – "Ist die Trägheit eines Körpers von seinem Energieinhalt abhängig?" ("Does the Inertia of a Body Depend upon its Energy Content?") – where he presents it, but in a rather different form. He is here more interested in the mass than in the energy. The equation then looked as follows:

$$K_0 - K_1 = L / V^{2*} v^2 / 2$$

where K is the kinetic energy, L the total energy, V the speed of light (later signified by c) and v is the speed between two coordinate systems. The expression is thus rather different from the modern form, but in words Einstein spells out the meaning which we see in the modern equation: "Die Masse eines Körpers ist ein Mass für dessen Energieinhalt", i.e. the mass of a body is a measure of its energy content.

The Best Known Physical Equation in the World: 2. Its History

Equals, =

For a long time equations were mostly written out in words. Most well-known symbols, including "=" became established during the eighteenth century.

$$E \quad = \quad m \cdot c^2$$

Energy

Energy became an important concept near 1850. Before then one mostly spoke of force. Energy is usually symbolized by W rather than E, but E is used in this equation since it is easier to understand.

At the middle of the nineteenth century was also the time that the first law of thermodynamics was first formulated (by several persons, independent of each other), $\Delta W = 0$ or $\Delta E = 0$. This means that the total energy (in a closed system) cannot be created or destroyed, only changed between different forms of energy, such as kinetic, potential, heat, radiation, etc.

Even today the main theory is that when the universe began (the Big Bang) 13.8 billion years ago, its total energy still remains zero, that is, it consists of equal amount of energy and negative energy.

Mass

The importance of mass (or weight) has always been known. Aristotle explained that massive objects want to return to their "natural place", the middle of the earth and the universe. When Copernicus showed that the earth was not still of the center of the universe (Aristarchos who argued for the same thing c. 280 BC had persuaded very few people) there arose a desparate need to explain mass (or weight). At least practically this was done by Newton, who showed that all massive bodies attract one another (and by what amount of force). However, he had no explanation for how this worked.

The Speed of Light Multiplied by Itself

"c" is an abbreviation of *celeritas*, Latin for fast, and began to be used as a sign for the speed of light in the late 1920s (as we saw in an earlier chart, Einstein simply used "v").

Many assumed that the speed of light was infinite, but Galilei considered an experiment with lantern on distant hills to determine its speed. That did not succeed, but already by the end of the seventeenth century the Danish astronomer Römer had calculated a good estimate for its speed, by using the light reflected on Jupiter's moons (compared to when the earth was close to Jupiter and when far away).

For further information, see Sander Bais, *The Equations: Ikons of Knowledge* (2005) and/or David Bodanis' book E=mc2 (2000).

History of Quantum Mechanics

The History and Development of Quantum Mechanics: 1. 1900–1927

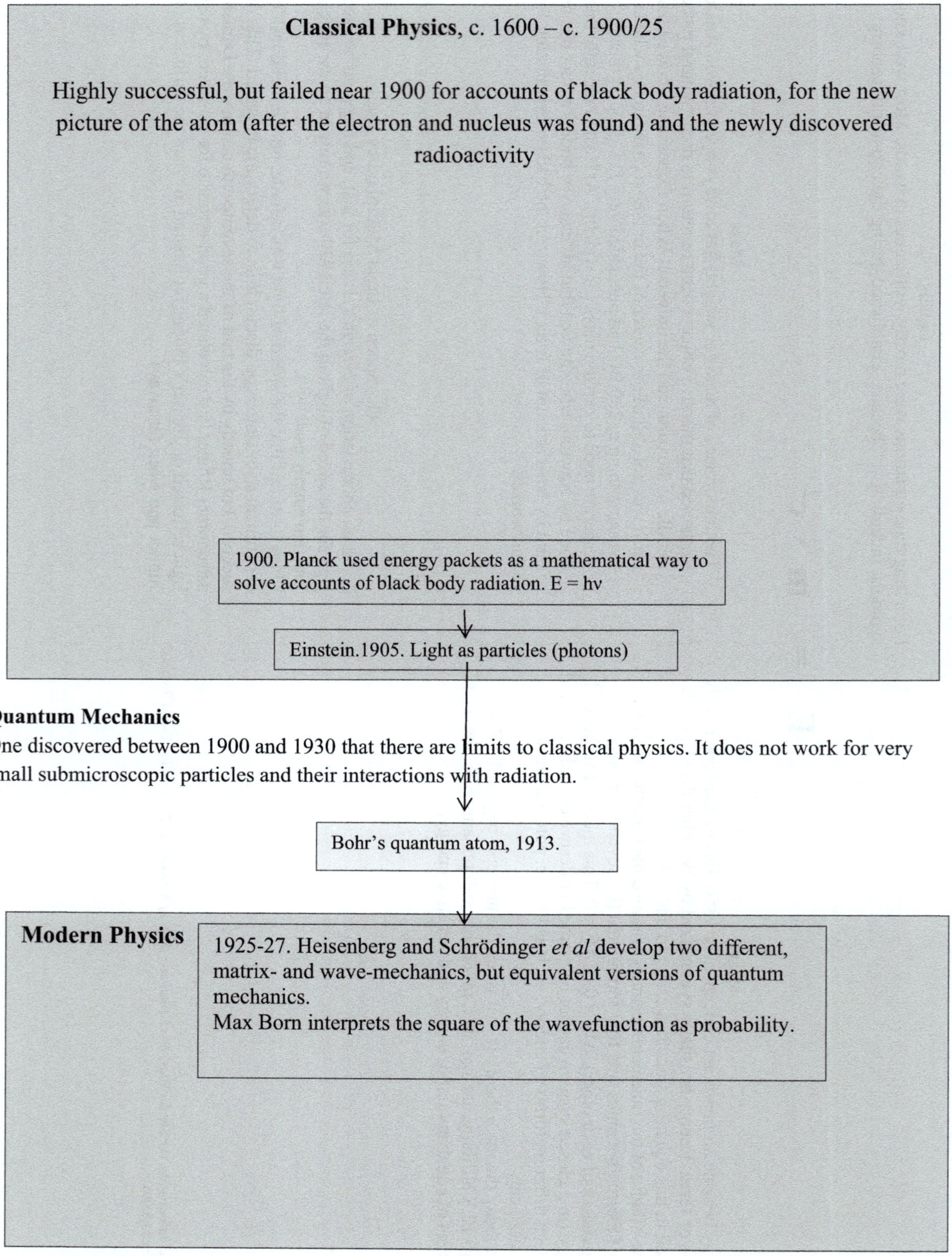

The History and Development of Quantum Mechanics (QM): 2. 1900–1927

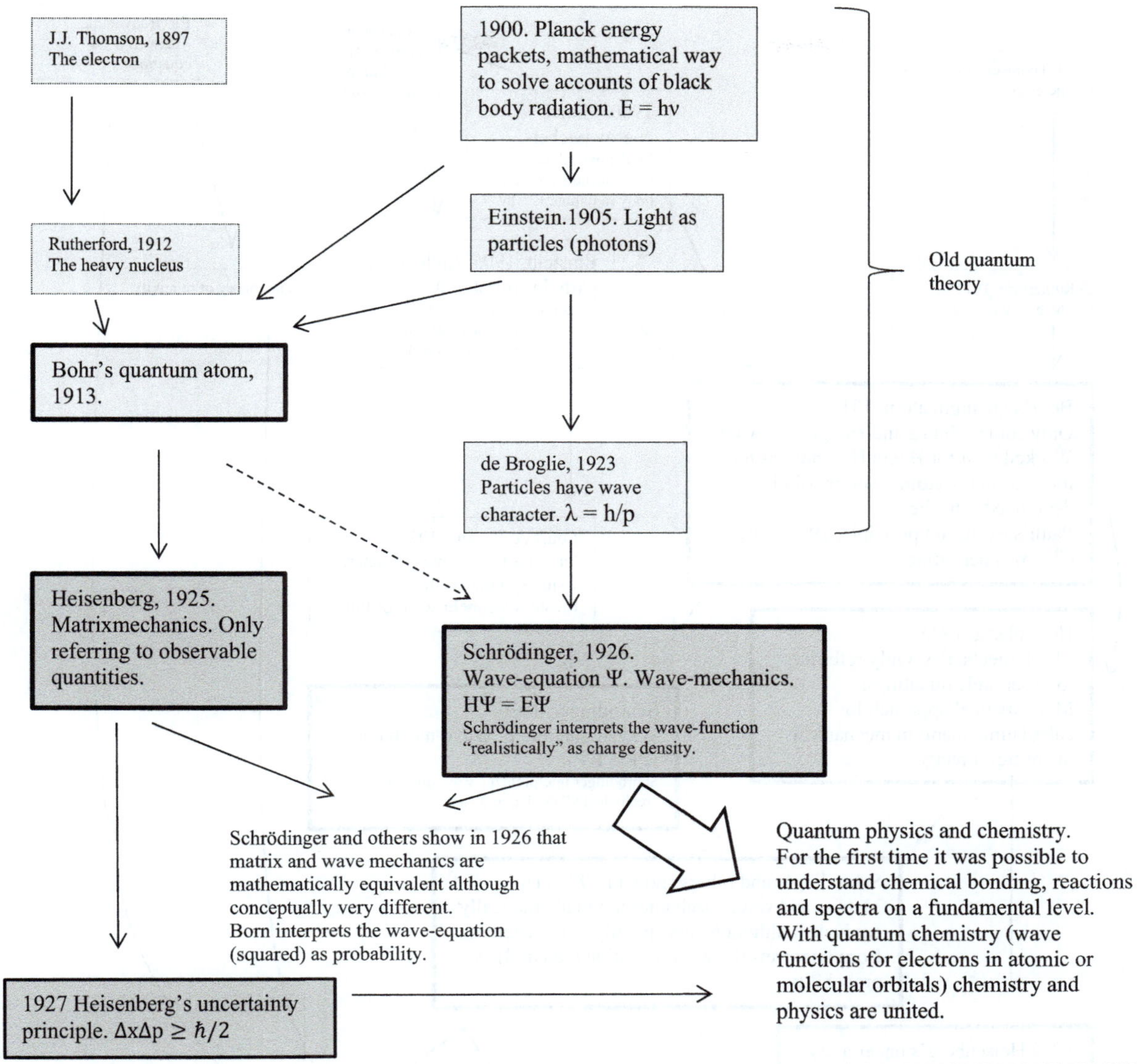

Interpretations of quantum mechanics:

Bohr and Heisenberg developed the Copenhagen interpretation, which is still the standard view of quantum mechanics.

Einstein, de Broglie, Schrödinger and later Bohm attempted to develop a "realistic" (non-probabilistic) and "hidden variable" interpretation, but have not convinced the majority of physicists.

After that quantum mechanics (QM) for atoms and molecules was basically developed, one moved on to new fundamental tasks:

1. QM needed to take special relativity theory into account (electrons travel fast).

2. QM (the physics of small charged particles interaction with each other) needed to be expanded to include how charged particles interact with electromagnetic fields, to construct quantum field theory (QFT).

3. One moved one step further, from atoms and molecules to the nucleus and elementary particles (and many new particles were discovered). The theoretical framework for this became known as the Standard Model.

The History and Development of Quantum Mechanics: 3. 1900–1932

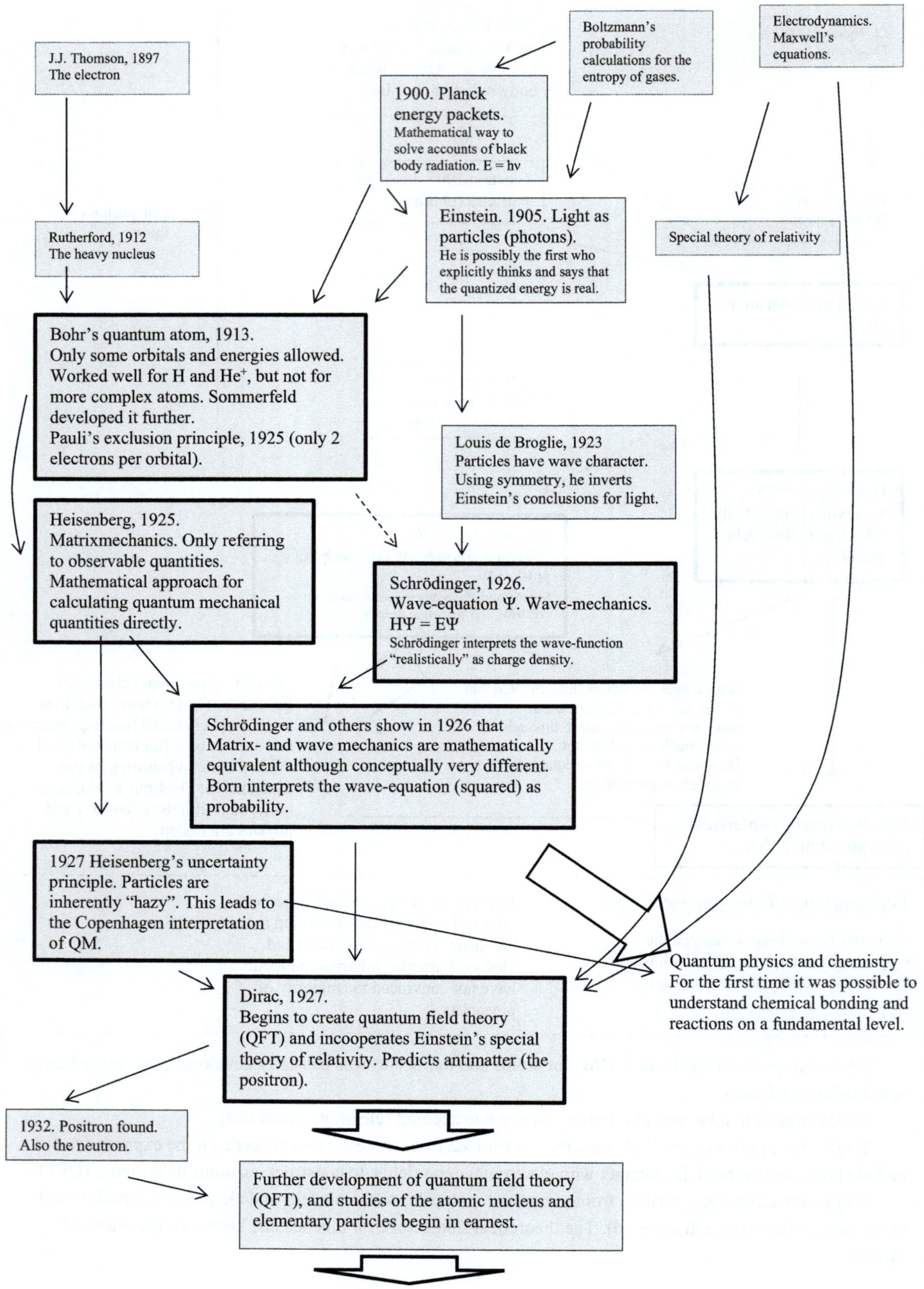

The History and Development of Quantum Mechanics: 4. 1900–2020 and Beyond

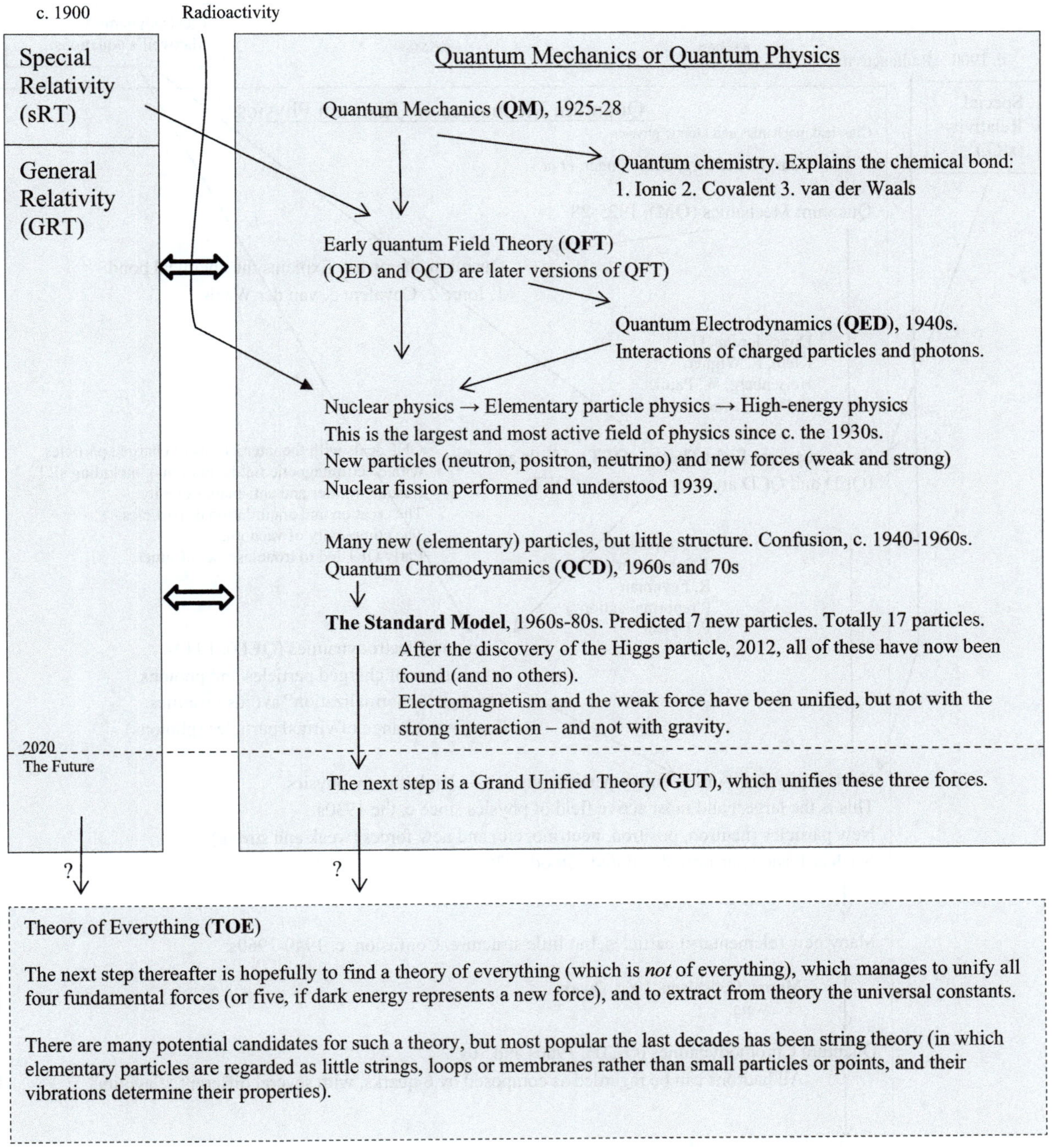

The Standard Model has been very successful, but is clearly not a final theory:

1. It has not unified the strong interaction with the electro-weak.
2. Most physicists believe that dark matter (c. 27% of everything in the universe), or large parts of it, consists of some other sort of elementary particles than those included in the Standard Model.
3. It cannot explain dark energy.
4. It cannot account for gravity.
5. It cannot account for neutrino oscillations and their non-zero mass.

The History and Development of Quantum Mechanics: 5. Expanded, with Names

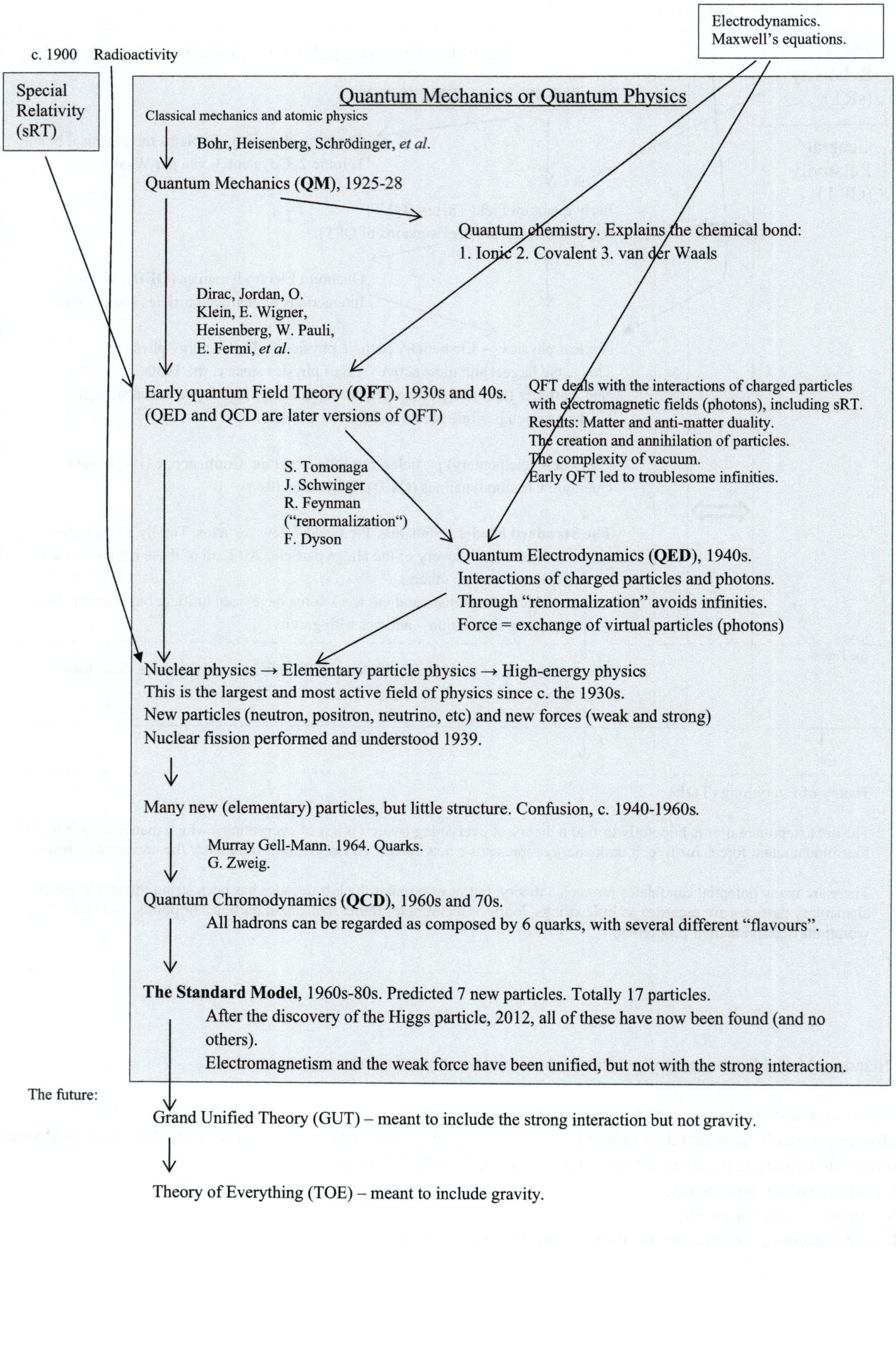

The State of Quantum Mechanics ("the Standard Model") Today

What Are the Fundamental Building Units in the Universe and How Many Sorts Are There?

According to the standard model, and according to all experiments done so far, "ordinary" matter consists of two kinds of elementary particles (leptons and quarks), and a third kind (bosons) are responsible for their interactions (forces).

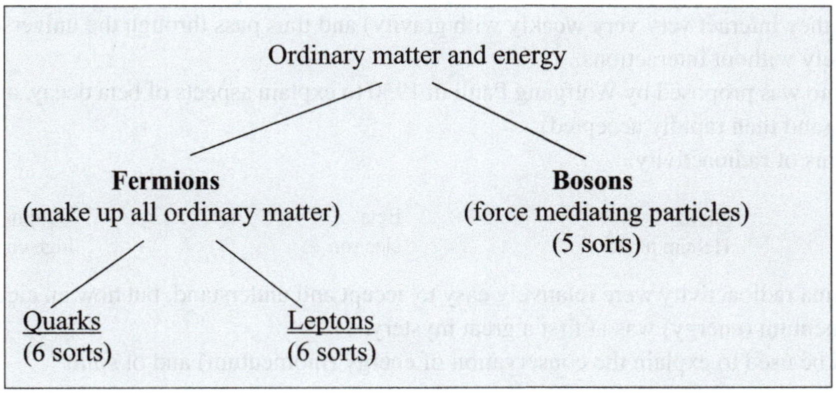

Together they can be added up to constitute a minimum of 17 sorts of particles at present, but there are several different ways to count them, so one can also get other numbers, with 61 as the maximum. In the 17, their anti-particles are not included, and if that is done, the number almost doubles (some particles, like light, are their own anti-particle, and would not increase the number), there are eight kinds of gluons and two kinds of W bosons, and the quarks can be distinguished by which of three "colors" they contain – the total number then becomes 61.

	Material particles (fermions)			**Force particles**
Quarks	Up (u)	Charm	Top	Photon
	Down (d)	Strange	Bottom	Z boson
				W boson
Leptons	Electron	Muon	Tau	Gluon
	Electron neutrino	Muon neutrino	Tau neutrino	Higgs

However, in the world around us today, only the first pair of quarks exists, up and down, the electron and the neotrinos (but the latter hardly interact with anything, and can, as a first approximation, be ignored). Thus there are basically only three matter particles (fermions) that we have to bother about, and which builds up everything around us, the up and down quarks and the electron. Proton = uud, neutron = udd.

However, since c. 2000 we know that "ordinary" matter only constitutes about 5% of all the mass-energy in the universe. The rest is called dark matter and dark energy, and most of dark matter is probably made of some other kinds of elementary particles.

The ingredients of the universe (according to the Planck satellite results):

Ordinary matter: 5%, dark matter: 27% and dark energy: 68%

The History of Our Knowledge of the Neutrino

The neutrino (the name means "the small electrically neutral one") is an extremely light electrically neutral elementary particle with spin ½ produced in nuclear reactions. They come in three forms, but originally only the electron neutrino was known. They are the second most common sort of (known) elementary particles in the universe, after photons, and they travel at close to the speed of light. They experience (interact only with) the short range weak interaction and gravity (but since they hardly have any mass they interact very very weakly with gravity) and thus pass through the universe (e.g. the earth or your body) almost completely without interactions.

The electron neutrino was proposed by Wolfgang Pauli in 1930 to explain aspects of beta decay, and further elaborated by Enrico Fermi in 1934 (and then rapidly accepted).

There are three forms of radioactivity:

Radioactivity:	Alpha	Beta	Gamma
Expelled particle:	Helium nuclei	electron	high energy photons

The alpha and gamma radioactivity were relatively easy to accept and understand, but how an electron could be expelled, and with varying momentum (energy) was at first a great mystery.

The neutrino could be used to explain the conservation of energy (momentum) and of spin.

$$n \rightarrow p^{+} + e^{-} + \nu_{e}$$

The electron neutrino (ν_e) was observed for the first time in 1956, the muon neutrino (ν_μ) in 1962 and the tau neutrino (ν_τ) in 2000.

Neutrinos from the sun were studied 1970–94, but the results did not agree with theory. Eventually one realized that the neutrinos are transformed into other forms of neutrinos.

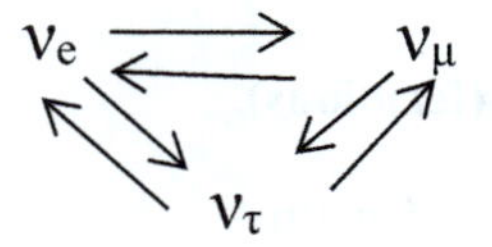

Just as there exists a cosmic microwave background (CMB) radiation, there also exists a cosmic neutrino background radiation.

The neutrinos are fantastic carriers of astronomical information, since they travel almost completely unaffected through anything and can tell about their origin unaffected.

The (newfound) fact that neutrinos have mass, however small, means that they are likely to account for some of the dark matter in the universe (although only a very small part of it).

The best theory we have at present for describing subatomic particles, the "standard model" fails to describe the neutrino (their mass and some of their behavior) – as well as accounting for dark matter and energy. It is likely that future work on the neutrino will be one of the best way to improve or to go beyond the standard model.

The History of Our Knowledge of Fundamental Forces

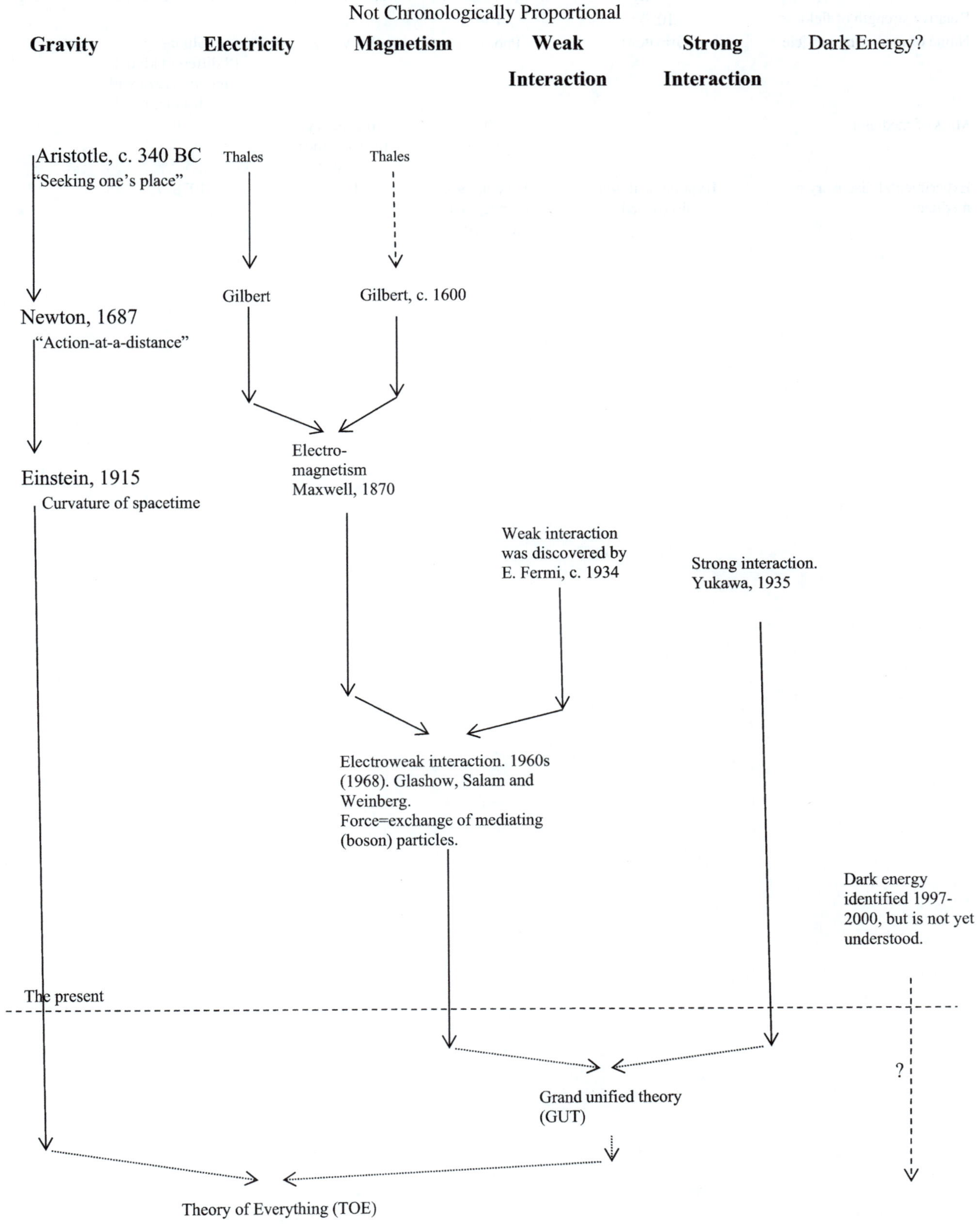

(The chart continues on the next page)

	Gravity	Electro-Magnetism	Weak Interaction	Strong Interaction	Dark Energy?
Relative strength of field	10^{-38}	10^{-2}	10^{-7}	1	?
Name of mediating particle	[Graviton]	Photon, γ	W^+, W^-, Z^0	Gluons, g (8 different kinds) (force *increases* with distance)	?
Mass of mediator	0	0	Yes, heavy (c. 100 times proton)	0	?, 0
Experimental discovery of mediator	Hypothetical, *not* discovered	Einstein, 1905 (conceptual discovery)	1983	1978–79	?

The History of the Views of Matter and the Building Blocks of the Universe

Not Chronologically Proportional

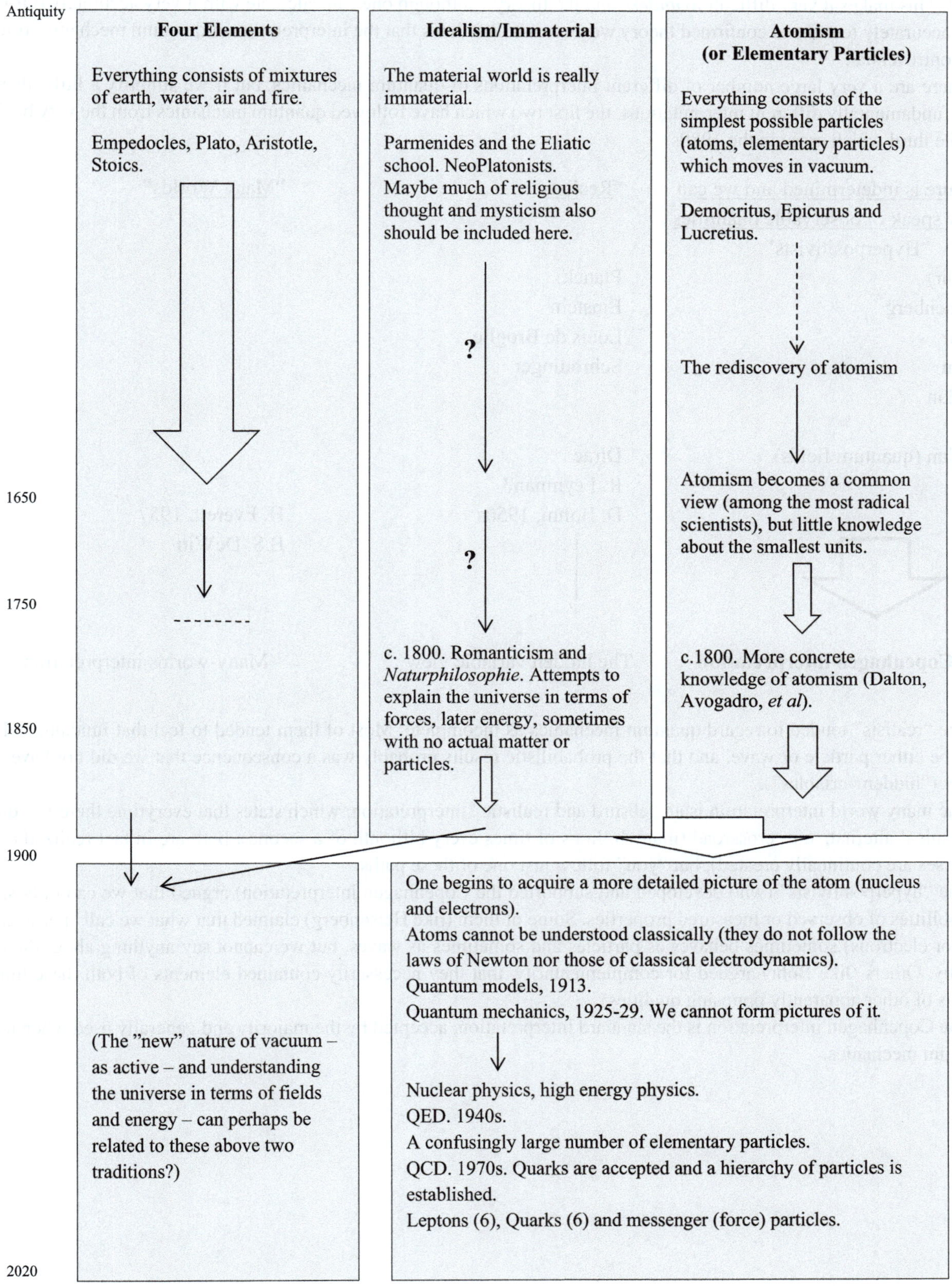

Three Main Philosophical Schools and Traditions Within Quantum Mechanics

Quantum mechanics is paradoxical for the human mind. The subatomic world appears to be very different from our ordinary world. This makes it very difficult to *understand* the theory – although one can calculate with it very accurately (QED is the most accurately tested and confirmed theory we possess). It follows that the interpretation of quantum mechanics is difficult and controversial.

There are a very large number of different interpretations of quantum mechanics, but if we simplify a little, there exist three fundamentally different interpretations, the first two which have followed quantum mechanics from the very beginning, and the third which arose in the 1980s.

Nature is indetermined and we can only speak of observable quantities "Hyperpositivists"	"Realists"	"Many Worlds"
(Bohr) Heisenberg Bohr Born Jordan	Planck Einstein Louis de Broglie Schrödinger	
Jordan (quantum fields)	Dirac R. Feynman? D. Bohm, 1950s	H. Everett, 1957 B.S. DeWitt
The Copenhagen interpretation	The hidden variable view	Many-worlds interpretation

The "realists" tended to regard quantum mechanics as incomplete. Most of them tended to feel that fundamental nature must be either particle or wave, and that the probabilistic results probably was a consequence that we did not have the full theory ("hidden variables").

The many world interpretation is an "absurd and realistic" interpretation, which states that everytime there is a quantum mechanical alternative or crossroad (many billions of times every billionth of a second), both are in fact realized (i.e. new universes are continually created), but "you" follow just one of these paths.

The "hyperpositivists" (who developed and supported the Copenhagen interpretation) argued that we can only speak of probabilities of observed or measured properties. Some of them (like Heisenberg) claimed that what we call "particles" (like light or electrons) sometimes behaves as particles and sometimes as waves, but we cannot say anything about their "real" identity. Others (like Bohr) argued for complementarity, that they necessarily contained elements of both these things (as well as of other apparently opposing qualities).

The Copenhagen interpretation is the standard interpretation, accepted by the majority and generally used when teaching quantum mechanics.

The Relation Between Classical Physics and Quantum Mechanics

Classical Physics (1600–1900)	Quantum Mechanics (Post-1925)
Valid for:	
Used to be believed to be valid for everything and everywhere. Nowadays, regarded as a good approximation for medium sized objects moving relatively slowly.	Extremely small objects (but also for larger objects). $h = 6.626 * 10^{-34}$ Js Quantum is Latin for 'unit' or 'small unit'.
Not valid for sub-microscopic size objects.	
Nature is and operates continuously. "*Natura non facit saltus*" (Nature makes no leaps).	Nature is and operates discontinuously (including probably also space and time, but that is not certain).
Continuous universe, energy and fields.	Quantized energy and fields.
Non-probabilistic. Theoretically exact and fully deterministic.	Inherently probabilistic (Heisenberg's uncertainty principle). Not fully deterministic (e.g. radioactivity).
Objects are either wave or particle.	Complementarity: objects are both wave and particle.
An object can only be at one place at the time.	An object seems to be able to be at more than one place at the time.
Reductionism possible.	Holism. An object cannot be observed without being disturbed – the observer needs to be included in the treatment.
Non-tunneling. A particle cannot penetrate high energy-barriers.	Tunneling. A particle can penetrate high energy-barriers.
Feels like it is basically possible to understand.	Impossible to understand – paradoxical (very different from our everyday world).
(No entanglement.)	Quantum entanglement seems to show that two entangled particles are "connected" over any distance so that when one is measured, the other's entangled property is also determined.

The Relation Between Relativity Theory and Quantum Mechanics

Relativity Theory	Quantum Mechanics
Valid for:	Valid for:
Large objects. Fast and slow moving objects	Extremely small objects (but also for larger ones)
Not valid for how sub-microscopic objects move	Not valid for speeds close to the speed of light
$c = 3*10^8$ m/s	$h = 6.626*10^{-34}$ Js

The two incompatible theories meet spectacularly in work on the Big Bang.
It is today thought that galaxies and galaxy clusters are the consequence of super-expanded quantum fluctuations.

Continuous field	Quantized field
Non-probabilistic	Probabilistic
Theoretically exact	Uncertainty
Difficult to understand, but not impossible. (different from our everyday world)	Impossible to understand – paradoxical (very different from our everyday world)

History of Modern Cosmology (The Big Bang Theory) (The Lamba-CDM Model)

History of Cosmology 1900–2020: 1. Outline

Until c. 1930, one had since antiquity regarded the universe as a whole as static.

Mostly Observations	Mostly Theoretical	Alternative Solution
	1917. Einstein's equations imply an unstable (expanding or contracting) universe, but he adds a "cosmological constant", Λ, to make it stable and static.	
1929. Hubble presents data which show that the redshift of galaxies (their velocity away from us) is proportional to distance.	1922-1932. Friedmann, Lemaitre, Eddington, Einstein and others solved the equations for an expanding universe, and combined with Hubble's observations accepted that the universe is expanding.	
	Lemaitre extrapolates backwards to a "primeval atom".	
	c. 1940. Hoyle and others can explain how stars work and how elements are created in them.	
	1948. Gamow and others calculate how several light elements arise in a hot big bang. Predict a cosmic microwave background (CMB) radiation. ↔	1948-70. Hoyle and Bondi propose an alternative explanation: "Steady state" cosmology. The universe is expanding and new matter continually created, but no big bang, and no CMB-radiation.
1965. Penzias and Wilson discovers CMB (2.7 K)		
	1980s. Guth and others propose inflation – very rapid expansion of space in early in Big Bang.	
1990-2013: COBE, WMAP, Planck satellite studies of the unevenness of the CMB. Conclusion: The universe $13.8*10^9$ years old, and consists of much dark matter and energy.		
c. 1997-2000. Studies of the redshifts of supernovas show that the expansion of the universe is not slowing down, but accelerating (due to dark energy).		

History of Cosmology 1900–2020: 2. More Details

At the beginning of 20th century, most astronomers assumed that the universe consisted of one large cluster (galaxy) of stars, with our sun possibly in the centre. Until c. 1930, one had since antiquity regarded the universe as a whole as static.

Mostly Observations	Mostly Theoretical
c. 1918–20. H. Shapley showed (by the use of cepheid-variable stars) that the cluster/galaxy was much larger than thought (the earlier view was c. 20 000 light-years across – much smaller than the present estimate of about 100–120 000ly), and that our sun is not at the centre, in the "Great Debate" in April 1920.	1917. Einstein's equations imply an unstable (expanding or contracting) universe, but he adds a "cosmological constant", Λ, to make it stable and static (to counter gravity).
c. 1925. E. Hubble shows that the spiral nebulae are independent entities far beyond our galaxy.	
1929. Hubble presents data that show that the redshift of galaxies (their velocity away from us) is proportional to distance.	1922–1932. Friedmann, de Sitter, Lemaitre, Eddington, Einstein and others solved the equations for an expanding universe, and combined with Hubble's observations accepted that the universe is expanding. Lemaitre extrapolates backwards to a "primeval atom" or "cosmic egg". Einstein takes away the cosmological constant from his equation for the universe (his "biggest blunder").
	c. 1940. Hoyle and others can explain how stars work and how elements heavier than helium are created in them.
	1948. Gamow and others calculate how several light elements arise in a hot big bang. Predict a cosmic microwave background (CMB) radiation.
1965. Penzias and Wilson discovers CMB (2.7 K)	
	1979–1980s. Guth and others propose inflation – very rapid expansion early in Big Bang. This explains several problems – the flatness of the universe, the "horizon problem" and the unevenness of the CMB (as due to expanded quantum fluctuations).
1990–2013: COBE, WMAP, Planck satellite studies of the unevenness of the CMB. Conclusion: The universe $13.8*10^9$ years old, and consists of much dark matter and energy.	During the 1990s a modified Big Bang theory developed into the standard theory – called Lambda-CDM model (Lambda for the cosmological constant and cold dark matter).
c. 1997–2000. Studies of the redshifts of supernovas show that the expansion of the universe is not slowing down, but accelerating (due to dark energy).	

Main Reasons to Believe in the Big Bang Theory and the Expanding Universe

There are many reasons why we believe that the Big Bang theory (the evolutionary theory of the cosmos) is basically correct. Many cosmologists argue that there are equally many and strong arguments, evidence and reasons to believe it to be fundamentally correct as there are reasons for accepting the theory of biological evolution (Darwinism). Some of the most important are (several of them have been mentioned in the above two charts):

1. The old theories for a stable universe turned out to be unstable.
2. Hubble's discovery of the systematic redshift of galaxies (interpreted as that they are moving away from us).
3. Einstein's equations predicted an expanding (or contracting) universe.
4. The probably strongest evidence is the cosmic microwave background (CMB) radiation, for which there is no other viable explanation, and which had to fit very well with the Big Bang theory (and had been predicted).
5. The estimated age of the universe by several other alternative methods than that of the BB (yielding 13.8 billion years) gives relatively similar results.
6. The correctly predicted relative amounts of H/He/Li in the universe.

The hypothesis of cosmic inflation, the exponential expansion of space during a short interval near 10^{-30} s after Big Bang, solves several problems with the Lambda-CDM–cosmology, but the inflation is still not well understood and not empirically verified.

The Size of the Visible Universe

It is difficult to get a reasonable concept and picture of astronomical distances, the size and structure of the visible universe. Kilometer is a useless unit for this purpose. Much better is to use the time it takes for light (300 000 km/s) to travel the distance. The moon is about one light-second away, the sun eight light-minutes, the nearest star 4.3 light-years (ly), the centre of the Milky Way 25 000 ly and the nearest galaxy, the Andromeda galaxy 2 million ly.

The Earth	Solar System	Nearest Stars	Milky Way	Galaxy Cluster The Local Cluster	Supercluster	Large Scale Sponge-structure	Visible Universe	?
Millilight-seconds	Light minutes	Light-years	100 000 ly	10 million ly	100 million ly	1000 million ly (=1 billion ly)	100 billion ly	?
Very approximate factor of increase in size:								
100 000	10 000	100 000	100	10	10	100		?

In the nineteenth and twentieth centuries the conceived size of the universe both grew and shrunk (in comparison to those who assumed an infinite universe).

The size of the visible universe is limited by the age of the universe and by the speed of light. At first one may think that the size (diameter) of the visible universe ought to be simply twice the age of the universe, in light-years, i.e. about 28 billion ly. However, the universe itself, space itself, is expanding and therefore the diameter of the visible universe is larger, about 50 billion ly. Assuming that the inflation hypothesis is correct, the actual universe is probably much larger than this. The size of the universe beyond the visible universe is not known. Furthermore, in the last decades it has become popular among astrophysicists and cosmologists to discuss or assume that there can exist many other universes than ours, that reality consists of a multiverse, although it is difficult to see how we could ever get any information to confirm or falsify this.

The State of Cosmology 2020: 1. Overview

Relative Speculative Theories about the Origin and the End of the Universe

Present Theories of the Origin of the Big Bang (or, what existed before the Big Bang?)

1. There was nothing before the Big Bang. Space, time, energy and matter suddenly appeared.
2. It is difficult (or impossible) to say how and why the Big Bang occurred. Ordinary space and time developed out of a primeval state possibly described by a quantum theory of gravity (but we do not possess any such quantum theory of gravity).
3. Our universe (and probably others?) were born out of another universe *ad infinitum*. Probably due to quantum fluctuations of vacuum. It is thus assumed that there exists many universes – the multiverse.
4. The Big Bang is only the latest stage in an ever cyclic universe, expanding and contracting, or just expanding and starting anew (e.g. the CCC cosmology of Roger Penrose or the ekpyrotic model of Neil Turok, Paul Steinhardt *et al*).

Until recently, the great majority of cosmologists only considered alternative 1. Recently, there has been much more attention than earlier directed especially at alternative 3, the multiverse alternative, although there is little (no) empirical evidence to distinguish the alternatives, nor is it clear how we even could confirm or falsify the existence of other universes.

The Future of the Universe according to Big Bang theories:

1. The acceleration (due to dark energy) ends and the universe continues to expand forever. The "heat death", i.e. everything eventually becomes "cold" and evenly spread out.

2. The acceleration continues or intensifies, and the universe will expand faster and faster, or even cause a "Big Rip", when all structure, even on the atomic level, is torn apart.

3. Acceleration changes to rapid deceleration and the universe collapses.

Before c. the year 2000 when the acceleration of the universe (dark energy) was discovered, a collapse of the universe due to gravity seemed a possible (fairly likely) scenario, but at present the third alternative is probably supported by the fewest number of cosmologists.

However, although the field of cosmology has moved from speculative hypotheses to a fairly exact science, there is too much we do not know, especially concerning dark energy, to scientifically speculate about the end of our universe.

The State of Cosmology 2020, and Its Results: 2

Much Less Speculative Account of the History of the Universe

The Development of the Universe According to the Big Bang Theory (the Lambda-CDM cosmology):

The start of the universe, 13.8 billion years ago.
(Time since the Big Bang)
10^{-43} s: "Planck time". The universe smaller than "Planck length", and we know nothing what happened before then.
10^{-38}-10^{-33} s: Inflation. The universe, spacetime (but not matter), expands much faster than the speed of light.
10^{-11} s: Matter becomes more prevalent than antimatter.
10^{-5} s: Protons and neutrons form from quarks.
0.01 – 300 s: Helium, lithium and heavy hydrogen nuclei form from protons and neutrons.

380 000 years: Universe cooled enough so that electrons can bind and form hydrogen and helium atoms. The cosmic microwave background (CMB) radiation released.

c. 100-300 million years: First stars and galaxies form (end of the dark period).

c. 0.5-1 billion years: The limit of current observations of stellar objects (maximum redshift). This is continually moved to earlier times. Several telescopes, including the James Webb, hope to reach the cosmic dawn.

The Milky Way galaxy forms about five billion years after the Big Bang (about 9 billion years before the present).

.

.

.

.

(Time before the present)
c. 4.6 billion years: The solar system forms.
c. 4.4 billion years: The earth is hit by large object and the moon is formed
c. 4.3 billion years: Ocean formation.

c. 3.8 billion years: Earliest life on earth?
c. 3.8 billion years. The accelerated expansion (dark energy) of the universe begins

c. 1.5 billion years: First eukaryotic cells. First sexual reproduction.

c. 0.6 billion years: First multicellular organisms

0.065 billion years: Dinosaurs go extinct, probably due to a large meteorite collision.
The present.

The History and Elusive Nature of Dark Matter

Our view of the cosmos has changed rather dramatically the last century, and even during the last few decades. As mentioned above, we know today that dark matter is much more prevalent in the universe than ordinary matter, although we know almost nothing about the nature of it.

In the late nineteenth century and early twentieth century Lord Kelvin and Henri Poincaré, responding to Kelvin, discussed the presence of dark unseen matter in the galaxy. In the early 1930s Fritz Zwicky discovered that the Coma cluster of galaxies seemed to contain far too little mass to prevent the galaxies from escaping the cluster. He called the missing mass dark matter. Thereafter this knowledge slowly developed until the 1970s when the existence of dark matter was further confirmed by Vera Rubin and Kent Ford, who discovered that many galaxies seem to contain only about 10 percent of the required mass for preventing stars from escaping. It was then still assumed that this was due to unseen ordinary (baryonic) matter.

Since then there has been much further confirmation of the presence of dark matter, by e.g. gravitational lensing, from the cosmic microwave background (CMB) radiation, and by observing x-rays from the motion of cosmic gas. Perhaps the most important observation has been that in the Bullet cluster, consisting of two merging galaxy clusters, dark matter seems to be partly separated from ordinary matter. It is today assumed that every galaxy has its own bubble of dark matter – a dark matter halo. Today most physicists assume that there simply is not enough ordinary matter (gas, brown stars, etc) to explain the missing matter, which instead is assumed to consist of some sort of exotic matter, that almost only interacts with ordinary, baryonic, matter through gravity.

However, there is almost no understanding of what dark matter is, although it is about five times as prevalent in the universe as ordinary matter. The main candidates of dark matter discussed at present are (not mentioned below are the so far unsuccessful attempts at modifying Einstein's general theory of relativity, e.g. MOND, modified Newtonian dynamics, and the possibility that dark matter consists of stellar mass black holes):

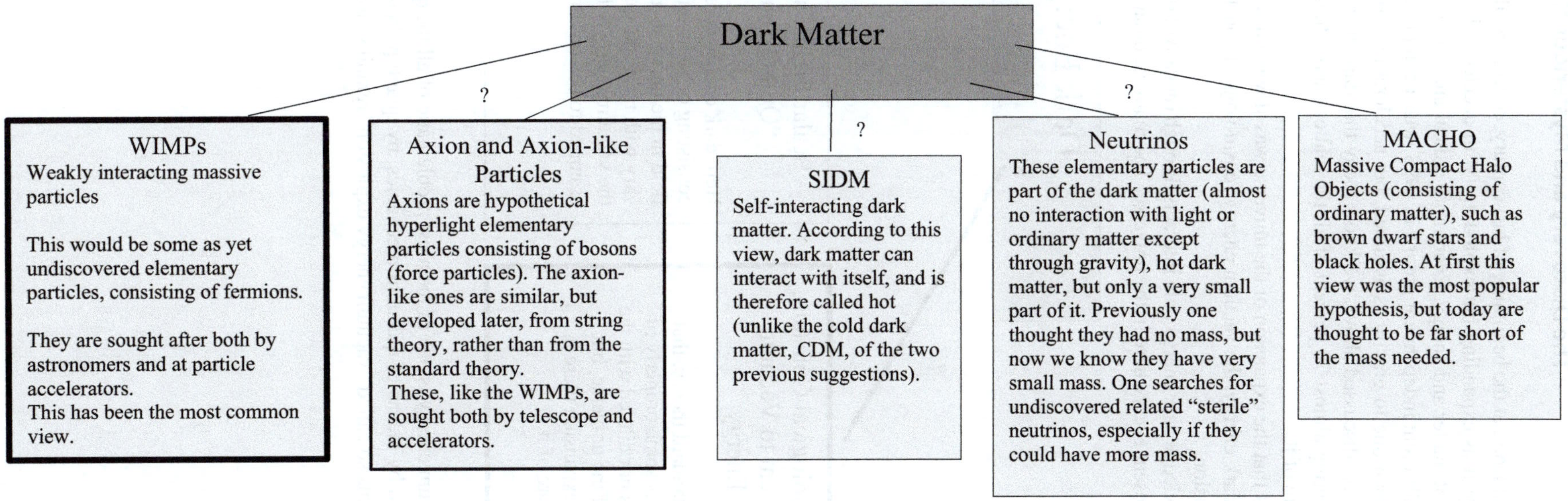

The Discovery and Mysterious Nature of Dark Energy

We have moved from, at the beginning of last century, conceiving the universe as static and consisting of ordinary matter, to not only seeing it as expanding but exploding (with accelerated expansion) consisting of mostly other, for us, unknown, "dark", forms of matter and energy. Even more radical and mysterious than dark matter is dark energy. Its discovery, published by two different independent groups in 1998, was a complete surprise for almost all. The expansion of the universe is not slowing down due to gravity as assumed, but accelerating. Some unknown energy, called dark energy, is pushing the universe apart. As discussed above, we know today that dark energy is much more prevalent in the universe than ordinary matter (constituting almost 70 percent of all the matter-energy there is in the universe), although we know almost nothing about the nature of it.

It is thought that the expansion of the universe was slowing down due to gravity for the first five to ten billion years, but thereafter the dark energy, the vacuum energy (cosmological constant) began to take over and, at first, slowly increase the speed of expansion.

To our knowledge dark energy only interacts through (negative) gravitation (or negative pressure). It seems to be extremely rarified, but present everywhere in the universe. The main candidates of dark energy discussed at present are the following:

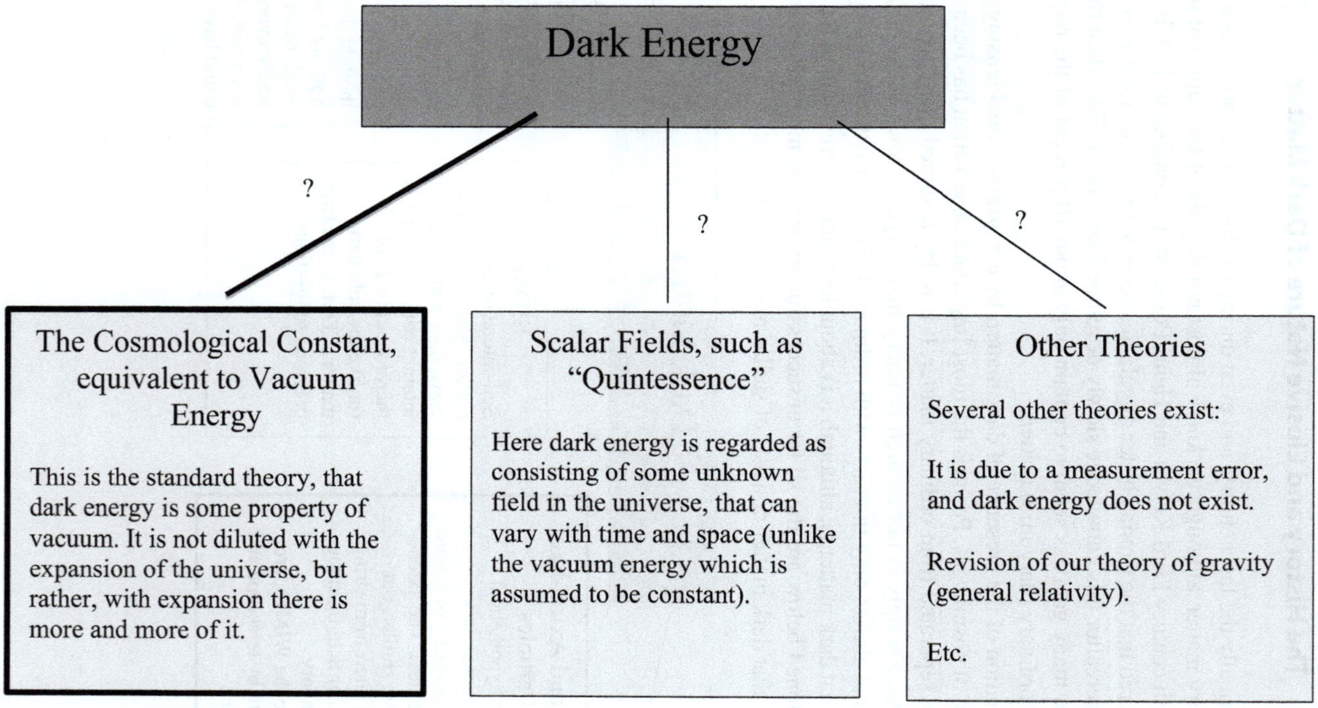

However, vacuum energy is one of the most problematic of all the questions of contemporary physics. When it is attempted to be calculated by means of quantum mechanics, by quantum field theory, one gets infinite or extremely high, obviously, false values. This so called "vacuum energy explosion" remains an unsolved problem in modern physics.

The Present Standard Model of Cosmology: A Version of Big Bang Theory

Lambda-CDM (ΛCDM) Cosmology

The Lambda-CDM (ΛCDM, Lambda Cold Dark Matter) cosmology is the simplest model of Big Bang cosmology that gives a reasonably good account of most of the known facts about the universe. It is based on that the universe began in a hot dense state, and the assumption that all the matter-energy consists of three constituents; ordinary (baryonic) matter – and the four forces associated with that: gravity, electromagnetism, weak and strong forces, dark matter and dark energy (the cosmological constant, lambda). It assumes that Einstein's general relativity is correct, and is named after its two main constituents, vacuum energy (lambda) and cold dark matter (cold, meaning that it moves significantly below the speed of light, and dark matter, almost certainly some form of matter unknown to us – probably consisting of some so far unknown forms of elementary particles). It is sometimes also called the concordance model, since it was a synthesis of earlier models. It requires six independent parameters. It was developed during the 1990s, out of earlier Big Bang cosmologies. Other more complex versions of Big Bang cosmology, needing more parameters, have been constructed, but have not so far been more successful.

Cosmological inflation – the ultrarapid expansion of space for a brief moment during the first second in early Big Bang, transforming the universe from about the size of an atom to that of a grapefruit – is not part of the model, but almost always added to it, since it solves a number of problems.

The model has been successful in accounting for many observations and made a number of successful predictions. Nonetheless, it also has a number of limits and challenges – apart from being based on the two incompatible or non-unified theories of general relativity and quantum mechanics – and many cosmologists assume that the model will need to be superseded by a better model. Among the challenges are the failure to understand or explain dark matter and dark energy, the Hubble tension (the fact that different methods of determining the Hubble constant gives somewhat different results) and the lithium problem (the observed amount of lithium in the universe seems to be several times less than predicted by the model).

The six independent parameters that need to be added to the model, and determined from observations, include the age of the universe ($13.799*10^9$ years), the density of ordinary (baryonic) matter and of dark matter. The other three parameters are less obvious and comprehensible for non-specialists.

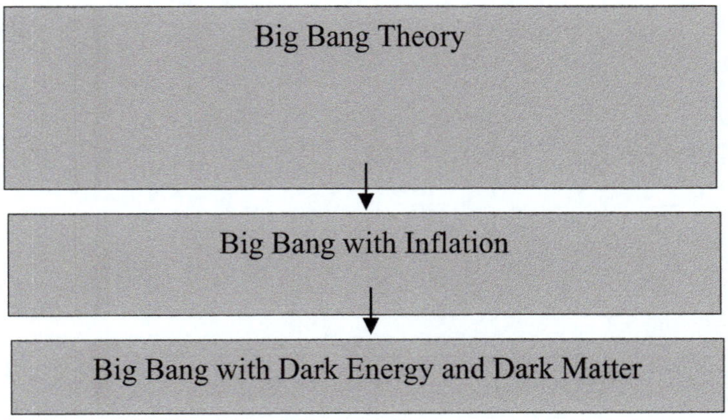

Gravitational Wave Astronomy: A New Window to the Universe

Gravitational waves – vibrations of spacetime – was predicted by Einstein in 1916, but thought to be too small to ever be observed.

Indirect observations of gravitational waves were made for the first time in the late 1980s by studying the slow changes in the orbit of a binary pulsar.

Direct observations of gravitational waves (GW) were announced in 2016 (and was awarded the Noble Prize in physics in 2017) of two colliding black holes. The waves are incredibly small, less than a hundred width of a proton. In the following years the sensitivity of LIGO has been improved and new observatories installed, leading to many new observations of colliding black holes and a few neutron stars collisions, with about one observation per week (when the observatories have been working).

This is a new window which will lead to much better understanding of the universe, especially concerning neutron stars, supernovas, black holes, possibly new so far unknown massive objects and about the early phase of Big Bang, possibly even about cosmic inflation (for which we otherwise have no direct evidence).

Gravitational waves, just like the case with studies by means of electromagnetic radiation, from radiowaves to gamma rays, need different kinds of observatories for different parts of the spectrum to be observed – and several such observatories are in the process of being constructed and tested. These will in the near future be significantly more sensitive and hopefully be able to observe a much broader range of gravitational wave incidents.

The cosmic microwave background (CMB) radiation, at present the most interesting and fruitful part of astronomy and cosmology, goes back until about 380 000 years after the Big Bang, while if we in the future are able to study the cosmic *gravitational* wave background (CGB) that can give us information about the Big Bang from the very beginning.

Two massive bodies (e.g. black holes or neutron stars)

give rise to gravitational waves when they spiral in and collide.

The Three Main Physical Theories of Reality: The Future of Physics

Three main theories or models stand at the forefront of our understanding of physical reality. Each theory is impressive and can explain much in great detail. The main problem is that general relativity is incompatible with quantum mechanics and the standard model, and cosmology is of necessity built on both of them. Special emphasis in this chart is on the limitations of these theories and on the need to extend them.

Gravity **General Relativity (GR)**	**Quantum Mechanics (QM)** **The Standard Model**	**Modern Cosmology** **The ΛCDM model**
Successes: It has been tested many times and always shown to be reliable. It competes with quantum mechanics for being the most accurate theory we possess.	Successes: Quantum mechanics and the standard model have been tested many times and shown to be reliable. The standard model contains the most accurate theory we have. It has predicted 7 elementary particles, all found – and no other.	Successes: We have a cosmology that can explain much of the development of the universe from the first second, even from much earlier. However, much remains unexplained, even highly fundamental things like dark energy, dark matter and the possible inflation.
Need for Extension: No sign of failure, but general relativity is a classical, non-quantized theory, and most researchers believe that it at very short distances must be quantized (and hopefully be unified with the three other forces). Many theoretical physicists search for quantum gravity.	Need for Extension: Dark energy. What is it? Vacuum energy? Dark matter. What is it? Neutrinos. That they have mass go beyond the standard model. Baryon asymmetry (matter-antimatter). Why and how? The magnetic monopole problem Vacuum energy explosion	Need for Extension: Dark energy Dark matter Neutrinos. We need a better theory Baryon asymmetry (matter-antimatter). Why is there much more matter than antimatter? The magnetic monopole problem Vacuum energy explosion The flatness problem The horizon problem Better understanding of inflation The amount of lithium The Hubble constant discrepancy, the "Hubble tension" We lack satisfying understanding of the initial conditions of the BB
Main Field of Research or New Development: Black holes Gravitational waves Testing gravity at short distances	Main Field of Research or New Development: Neutrinos Dark matter The search for new elementary particles with accelerators	

Possibly the future?

A possible new theory called "Theory of Everything"

History of the Life and Earth Sciences

Biology and Geology

7

T. H. Brobjer, *Atlas of the History of Modern Science 1500-2020*, https://doi.org/10.1007/978-3-031-76627-5_7

Map of the History of Biology (Natural History) and Medicine: 1. Outline

Chronologically Proportional

There were primarily two traditions from antiquity and the Middle Ages that influenced and constituted the main fields of biology; natural history (in which Aristotle and Theophrastus wrote extensively) and medicine. These two fields met around 1600 in the field of botany, which during the period 1400–1800 was the most central field of biology.

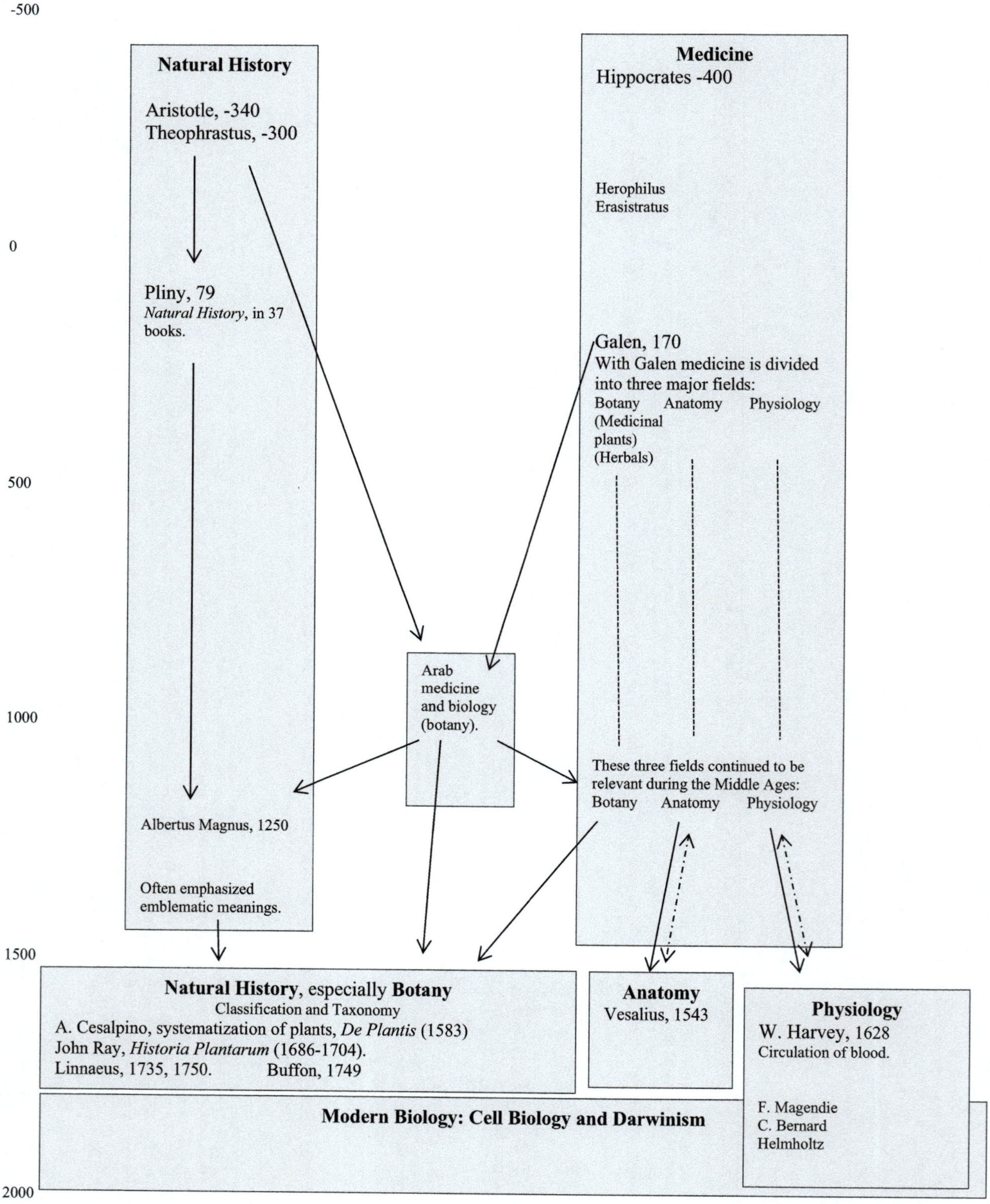

History of Biology (the Science of Life), Antiquity – 1800: 2. Overview

Not Chronologically Proportional

There were primarily two traditions from antiquity and the Middle Ages that influenced and constituted the main fields of biology; natural history (in which Aristotle and Theophrastus wrote extensively) and medicine. These two fields met around 1600 in the field of botany, which was during the period 1400-1800 the most central field of biology.

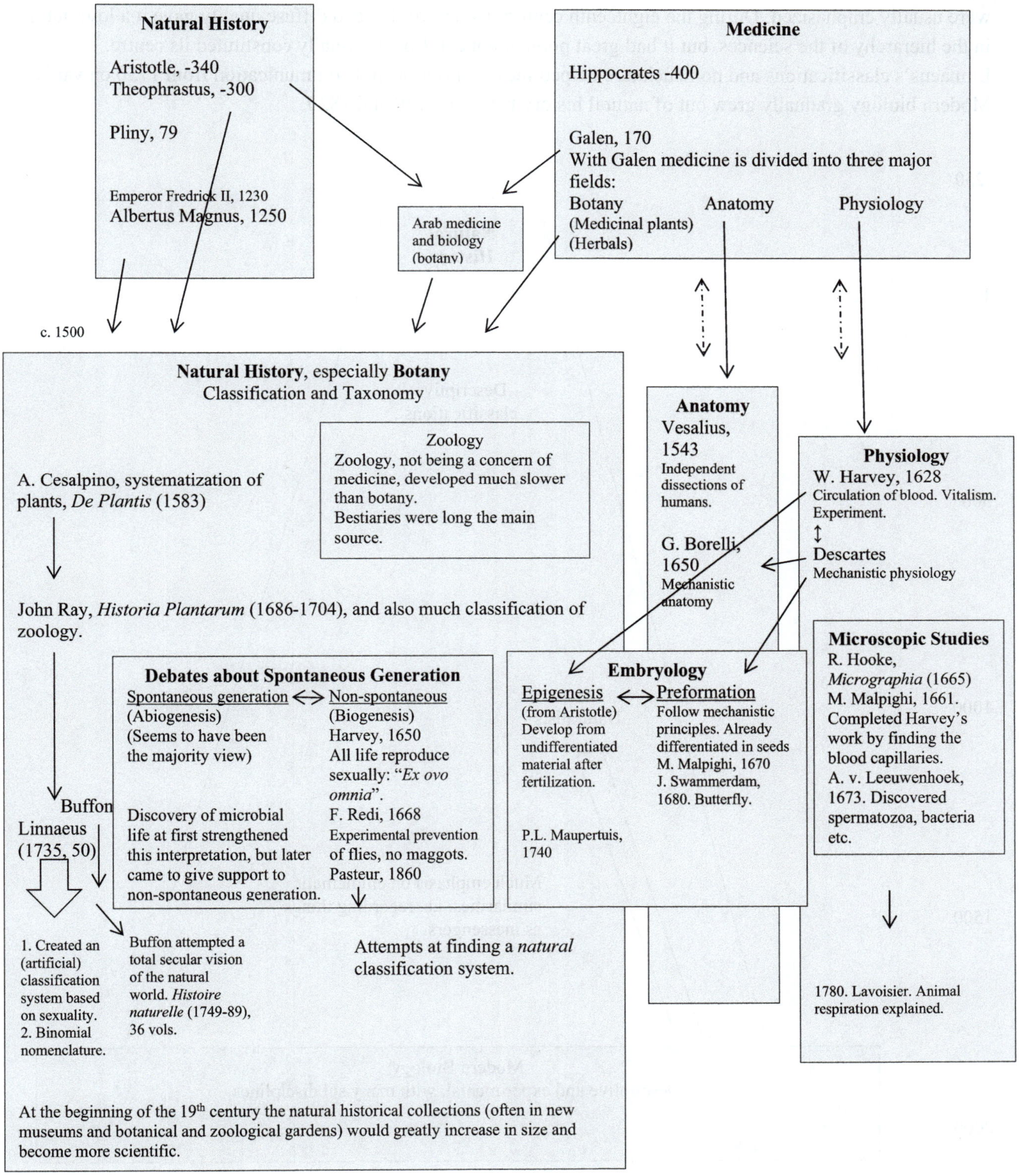

The History of Natural History: 1. Overview

Natural history was originally the study of the whole natural world, including what is now called meteorology, geology, geography and much else. Its nature was always descriptive. It was distinguished from natural philosophy, which was more theoretical and emphasized laws and generalities. Both of them were often closely linked with natural theology. Moral and didactive aspects were often emphasized. During the Middle Ages and the Renaissance emblematic meanings (based on assumed similarities with human body parts etc) were usually emphasized. During the eighteenth century its descriptive and diffuse quality gave it a low status in the hierarchy of the sciences, but it had great popular appeal. Botany usually constituted its centre. Linnaeus's classifications and nomenclature helped the organisation and communication from 1750 onwards. Modern biology gradually grew out of natural history between 1750 and 1820.

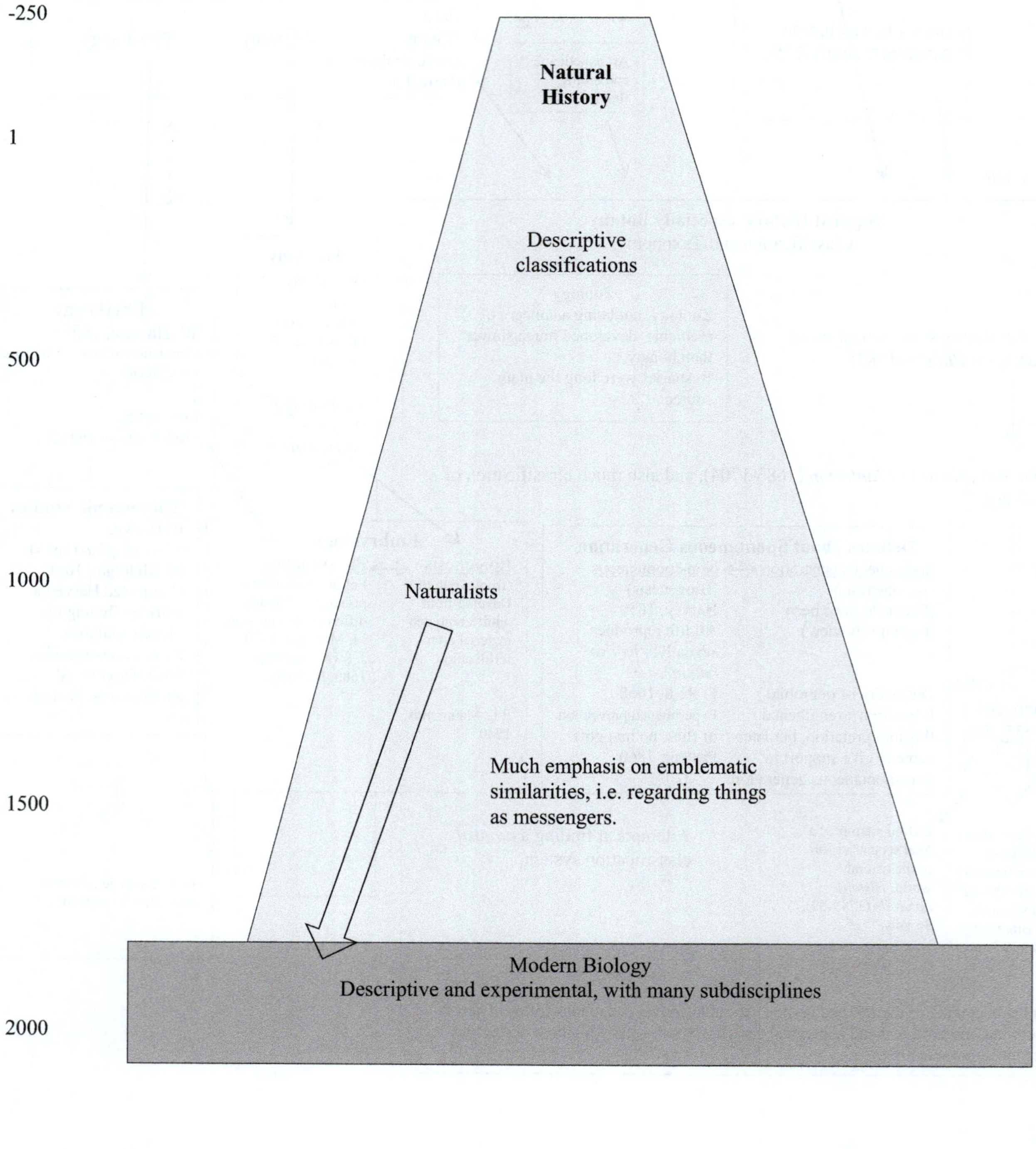

History of Biology: Vitalism and Teleology contra Mechanism: 1. Overview

Life and living matter seem very different from dead matter. One way to account for this difference is vitalism, that life is based on a special force, soul, substance or other property. Others argued that life is just particular arrangements of ordinary matter (usually called mechanism). Life also often seem to be able to act with purpose (teleology), i.e. not only be dependent on physical causality, cause and effect, so to speak "from below", but also by purpose (*telos*) "from above" (goal-oriented). These apparently special characteristics of life were perhaps most visible and mysterious in reproduction, and the attempt to understand this was prominent during the eighteenth and nineteenth centuries. This chart shows a simplified view of the main paradigms.

However, one should be aware of that the distinction between life and dead matter was much less clear before c. 1800 than thereafter (due to belief in spirits, in that stones etc could come alive, in the afterlife, in spontaneous generation etc was common).

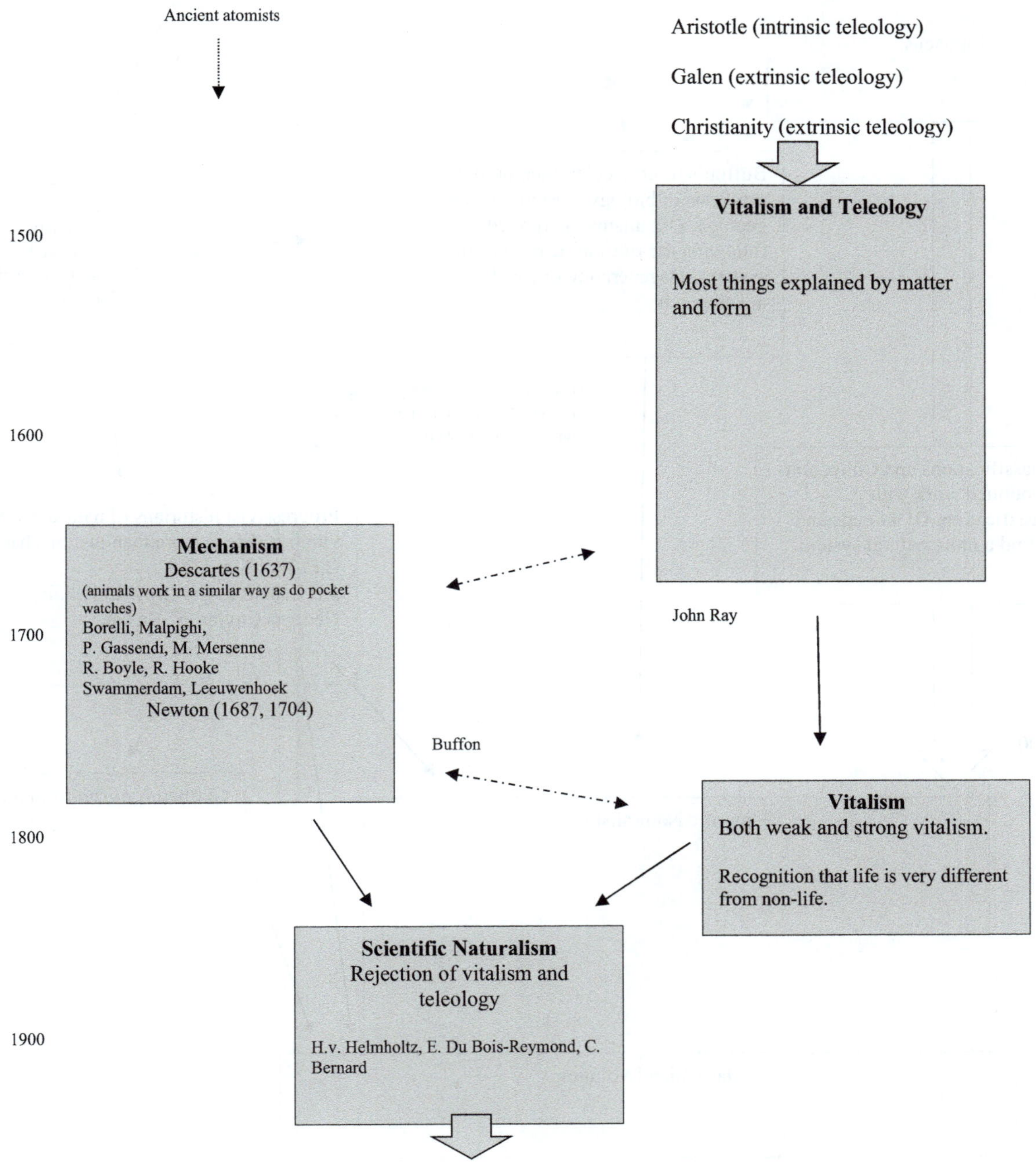

Main Themes of Eighteenth Century Natural History

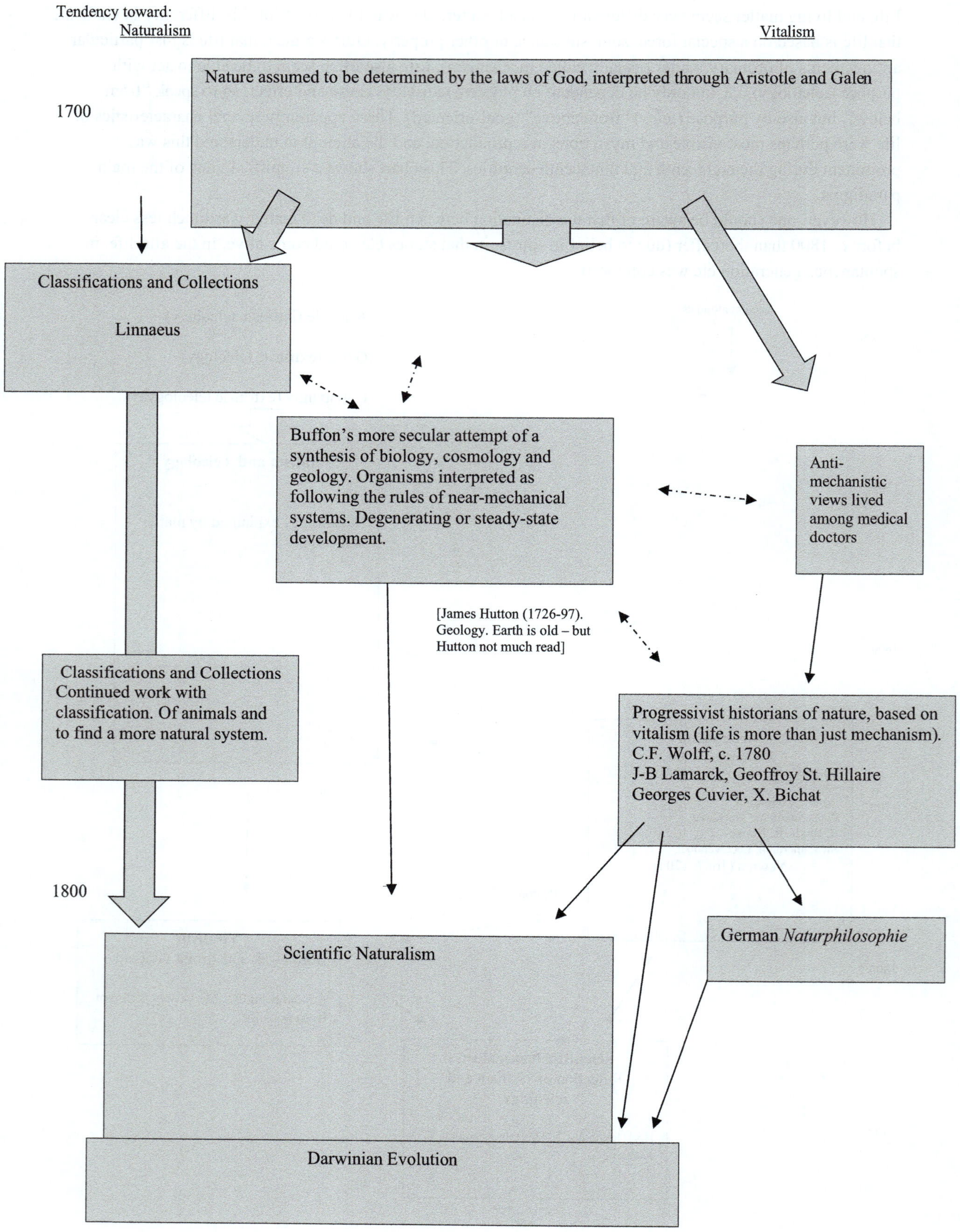

Important Eighteenth Century Natural Historians (Biologists)

Naturalists

[N. Grew, 1682. Sexuality in plants.]

R.J. Camerarius, S. Vaillant 1717
Plants reproduce sexually

Gilbert White, 1789
Selborne

Biogeographers

J. R. Forster
E. Zimmermann
A. v. Humboldt

Natural History, Classification

[A. Cesalpino, 1560]

John Ray, *History of Plants* (1686, 88, 1704)

Linnaeus, 1740
Systema Naturae, 1735
Philosophia Botanica, 1751. Used reproductive part of plants to systematize them (artificial system). Assumed universality of sexual reproduction. Binomial system of nomenclature. Introduced man into the system.

The Linnaean Society of London founded 1788

[Steno, 1660s. Fossils are dead organisms.]

Buffon, 1750
Histoire naturelle, 1749-1804. Mechanistic explanation of reproduction. Critique of Linnaeus system as too artificial. Synthesis of cosmology, geology and biology. Earth is old.

Georges Cuvier
Comparative anatomist and paleologist. Possibly the most authoritative natural historian at the end of the century. Against species transformation

Microscopic Biology

[Hooke, *Micrographia*, 1675. The cell. Leeuwenhoek, 1676. Bacteria.]

A large number of biologically relevant microscopic studies made, by e.g. Réaumur, Ch. Bonnet, L. Spallanzani and B. Albinus.

Evolutionary Thought
(species transformation)
J-B Lamarck
E. Geoffroy St Hilaire

Others

Erasmus Darwin, 1770
Emphasized environmental factors. Suggests descent from common ancestor and the survival of the fittest. *Zoonomia* (1794).

L. Galvani, 1780s
"animal electricity"

Lavoisier, 1790
Respiration explained

History of Embryology, Generation, Reproduction 1650–1800: 1. Outline

Embryology, the study of the embryo and its development, was always a promising avenue for gaining knowledge and classification of organisms. This approach began already in antiquity (among others, by Aristotle, who examined the embryology of chicks), but was emphasized c. 1600-1900, especially 1650-1800, when embryology was one of the most important and active fields of biology (and medicine).

There were basically two schools for how embryo developed, "epigenesis", believing that undifferentiated material began to organize itself after fertilization, beginning already with Aristotle, and often teleological and vitalistic, and "preformation", a more mechanistic (anti-teleological) view beginning in the 17th century, and claiming that the fetus grows automatically according to the already existing copy in the sperm or egg.

General characteristics of the debate:

Preformation / Pre-existence		Epigenesis
(The fetus grows according to an already existing copy in the cell)	<-·-·-·->	(Material organizes itself after fertilization)
Supporters often held atomistic and mechanistic views.		Supporters often held animistic/vitalistic, or alternatively materialistic, views.
Within this tradition, there was a debate about where the "form" was placed, in the egg or in the sperm.		This view seemed to require a mystical "vital" force to form the matter during an embryo's growth.
The standard view became that God must have implanted these "forms" in the first parents of each species (*emboîtement*).		

With the development of genetics (and that inheritance is due to the chromosomes, nucleic acid and in 1953, to DNA) the debate died out (as a sort of compromise of both theories, but most would today argue that epigenesis was more right than preformation – but this could not be known at the time of the great debate).

History of Embryology, Generation, Reproduction: 2. Overview

Some of the most influential participants in the debate:

Preformation/Pre-existence

Greek atomists

↓

G. Fabrici (1621) and others investigated embryoes and argued that they build themselves from a bony framework.

Many scholars opposed Harvey's views, c. 1650-1800. Most influential was M. Malpighi (1673), whose arguments made preformation the majority view for over a hundred years. Some argued that preformation came from the egg (ovism), others from the sperm.
Swammerdam
Réaumur
Ch. Bonnet
Albrecht von Haller, irritability mechanist. c. 1760
L. Spallanzani

Epigenesis

Aristotle

↓

W. Harvey, *De geneative animalium* (1651) argued against his teacher Fabrici and postulated that all females produce eggs (ova) which could be stimulated to develop by male semen, which made it germinate and divide (epigenesis)

J. B. van Helmont;
Leibniz
E. Stahl
Linnaeus
Maupertuis; Buffon

C.F. Wolff, vitalistic epigenesist (1759)
J. F. Blumenbach (1781, "formative force")
Several embryologists influenced by *Naturphilosophie*, c. 1810-40.
E. von Baer

↓ ↓

The controversy regarding preformation and epigenesis continued, but weaker, in the 19th century. Preformation theories declined and epigenesist views dominated. Due to improved microscopes, new staining techniques, etc, many new discoveries were also made.

Naturphilosophie ⇨ 1817. Ch. H. Pander. Blastoderm developed into three separate layers and then into different organs.

1827. E. von Baer discovery of the egg in mammals.

1850-55. R. Remak showed that the germ layers consisted of cells (and that the egg was a cell).

1875-80. Fertilization consisting of an egg meeting a sperm, that fuse and form the embryo that begins to divide.

Recapitulation

Influenced by Drawinism, E. Haeckel formulates the biogenetic law, 1866, that ontology (the development of the individual) recapitulates phylogeny (the development of the lineage to which it belongs), determined by heredity and adaptation. This view was highly influential, but was heavily criticized during the end of the 19th century.

History of Modern Biology 1800–2000: 1. Overview

Modern biology can be said to have been born around 1800 – which is also when the word biology began to be used (by e.g. Treviranus, Lamarck, Comte) – due to work by French biologists (such as Buffon, Cuvier, Geoffroy de Saint Hilaire and Lamarck), better understanding of physiology (Magendie), German *Naturphilosophie* (Goethe and L. Oken) and the beginning of cytology (cell biology) and histology (the study of tissues). Until then, biology consisted mainly of natural history, basically consisting of taxonomy and anatomy. With the new biology around 1850 biology became experimental. The two main fields for the next hundred years were Darwinian evolution and cell biology. Evolution became the framework within which almost everything was viewed: "Nothing in biology makes sense except in the light of evolution," Theodosius Dobzhansky.

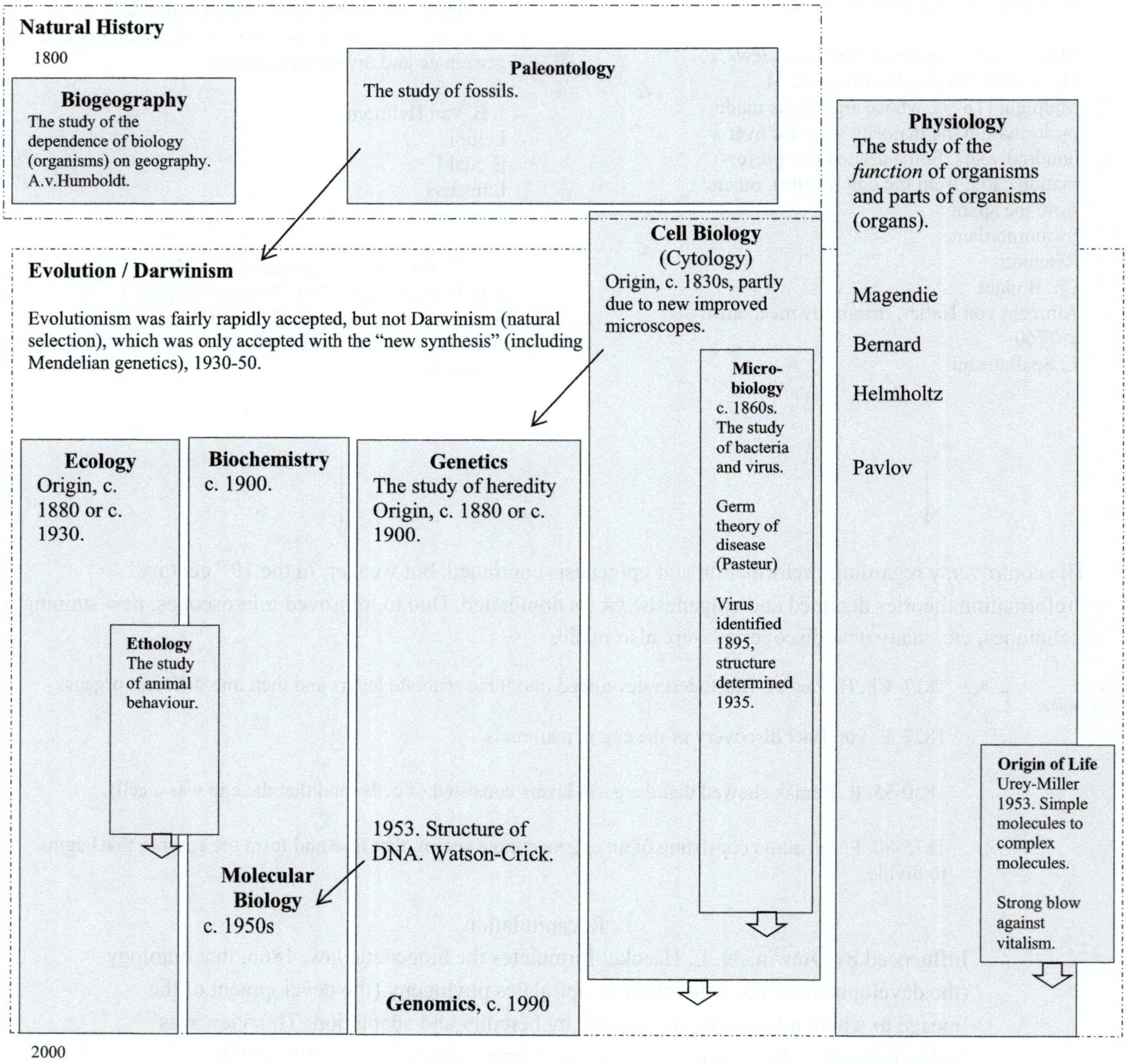

History of Modern Biology 1800–2000: 2. Overview

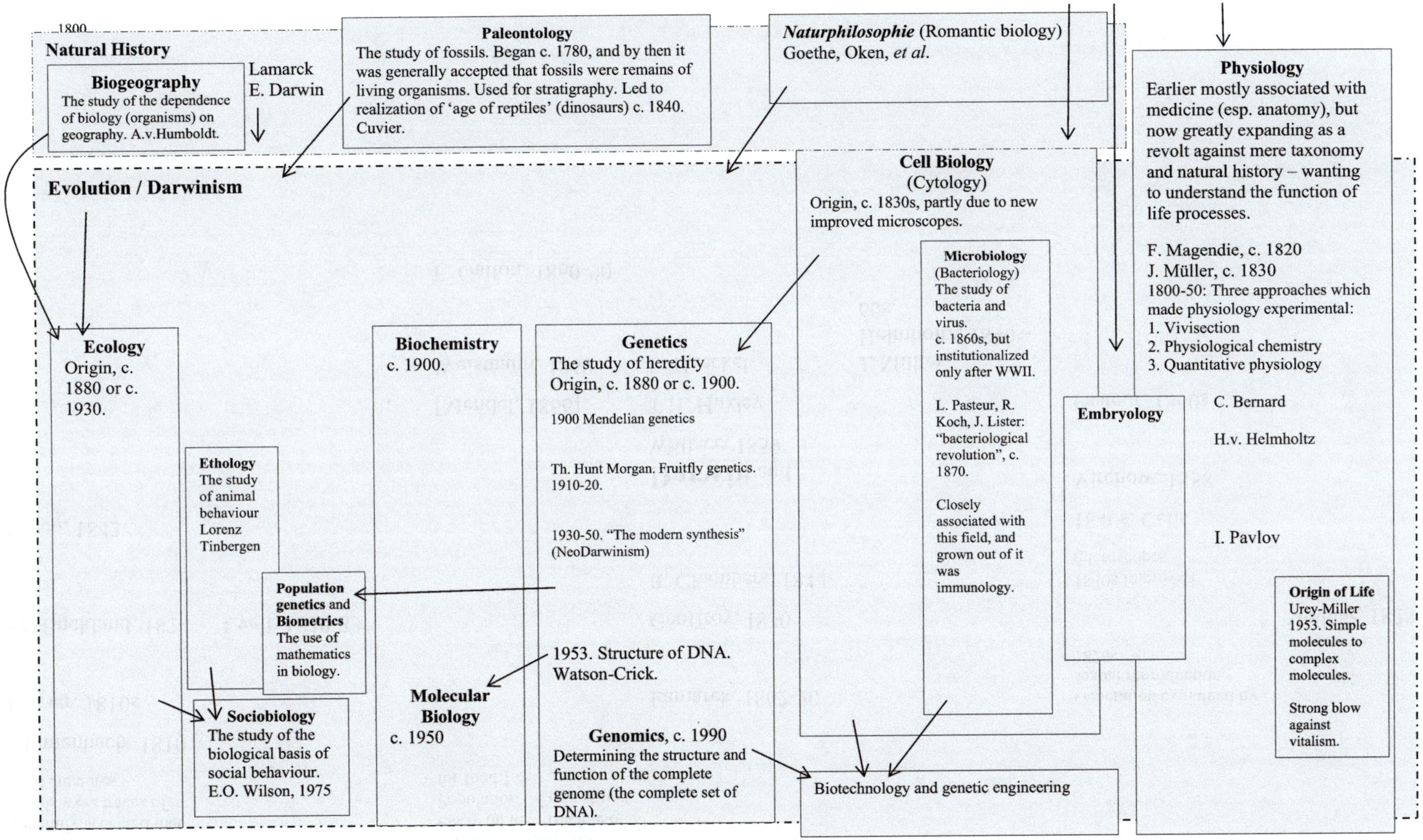

Table over the Most Important 19th Century Biologists: 1. Overview

Paleontology	Geology	Hereditary / Genetics	Natural History	Medicine/ Physiology	Laboratory Biology / Microbiology	Chemistry
By c. 1800 it was generally accepted that fossils were traces of dead creatures.		[Malthus, 1798, "An Essay on the Principle of Population". Competition for food.]				
Blumenbach, 1810						
Cuvier, 1810s			Lamarck, 1802-20		Generation explained by sexual reproduction, 1820s	
W. Buckland, 1824	Lyell, 1830-33		Geoffroy, 1830			Wöhler, 1828
			R. Chambers, 1844		1840s Improved microscopes	
Owen, 1842					1840s. Cells	
			Darwin and Wallace, 1859		Virchow, 1858	
		[Mendel, 1866]	T.H. Huxley		Pasteur, 1860s	
		Weismann, 1868	E. Haeckel	J. Müller, Helmholtz, 1840s-60s.		
		F. Galton, 1860-90				

Table over the Most Important 19th Century Biologists: 2. Overview with Summaries

Paleontology	Geology	Hereditary / Genetics	Natural History	Medicine/ Physiology	Laboratory Biology / Microbiology	Chemistry
By c. 1800 it was generally accepted that fossils were traces of dead creatures.		[Malthus, 1798, "An Essay on the Principle of Population". Competition for food.]				
Blumenbach, 1810 Many animals and plants have become extinct, and others created.						
Cuvier, 1810s The father of paleontology and comparative anatomy. Fossils can be used to date rock, stratigraphy. Species have become extinct, but do not change. Opposed Lamarck.			Lamarck, 1802-20 First systematic theory of organic development. Acquired characteristics are inherited. Teleological. *Zoological Philosophy* (1809).		Generation explained by sexual reproduction, 1820s Prevost, Dumas, Baer.	
W. Buckland, 1824 First paper describing a dinosaur.	Lyell, 1830-33 *Principles of Geology* Slowly evolving geology. Critical of Lamarck.		Geoffroy, 1830 Developed and circulated Lamarck's theory.			Wöhler, 1828. Synthesis of an organic compound (urea) from inorganic material.
			R. Chambers, 1844 *Vestiges of the Natural History of Creation*.		1840s Improved microscopes	
Owen, 1842 Coined the term dinosaur.					1840s. Cells The cell as the fundamental unit of life. Schleiden (plant cells), Schwann (animal cells).	
			Darwin and Wallace, 1859		Virchow, 1858 All cells from cells.	
		[Mendel, 1866] Discovers the laws of inheritance, but is not read. Rediscovered 1900.	T.H. Huxley Skillful biologist and Darwin's "bulldog".		Pasteur, 1860s Germ theory of disease. 1864 disproved spontaneous generation of life.	
		Weismann, 1868 Inheritance through "germ plasma" from sperm and egg. Denies that acquired characteristics are inherited.	E. Haeckel	J. Müller, Helmholtz, 1840s-60s. Life obeys mechanistic principles, can be reduced to the physical sciences.		
		F. Galton, 1860-90 Studied inheritance of measurable characteristics. *Natural Inheritance* (1889) used statistical methods. Eugenics.				

The Development of Biology in Terms of Its Most Important Subdivisions 1800–2000

The largest or most active field(s) in bold. Many “minor” fields are, of course, not listed.

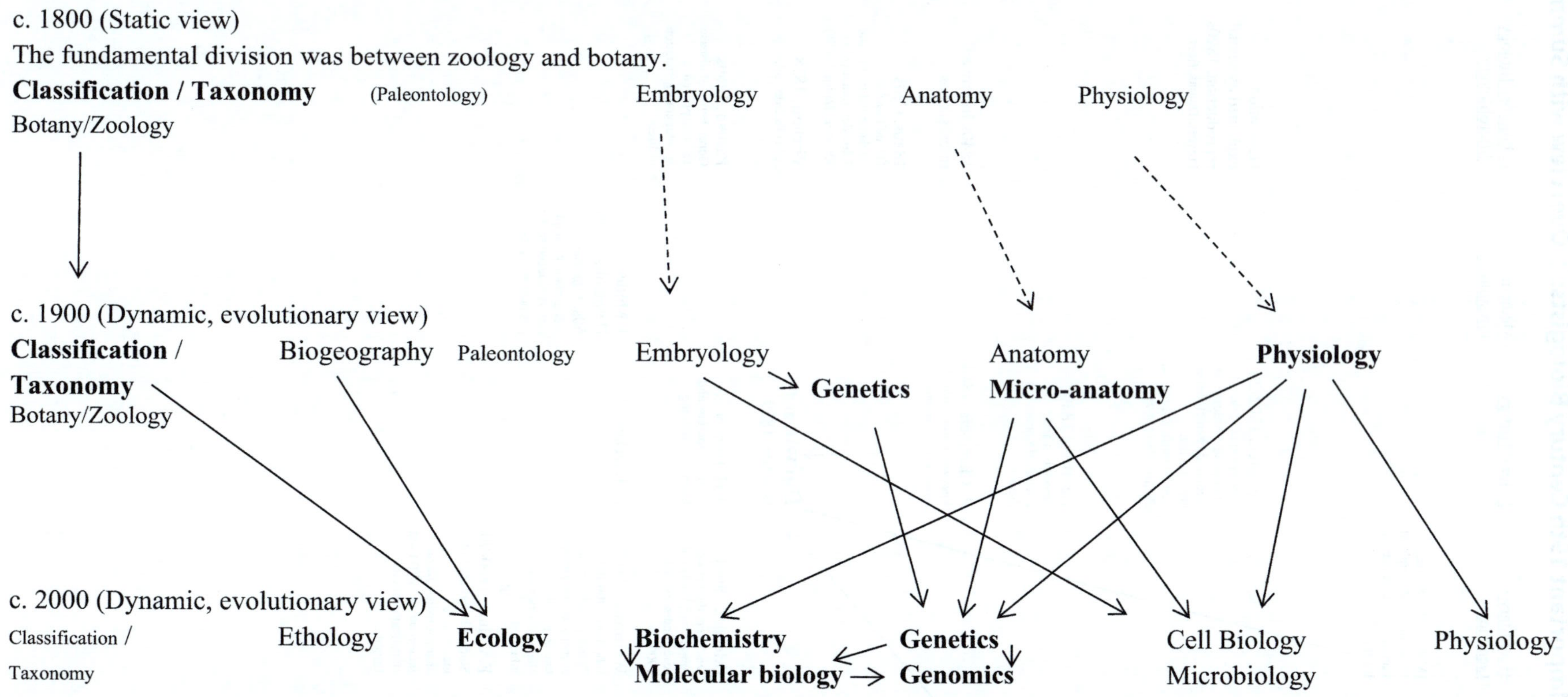

The History of Classifications and the Tree of Life 1780–2020

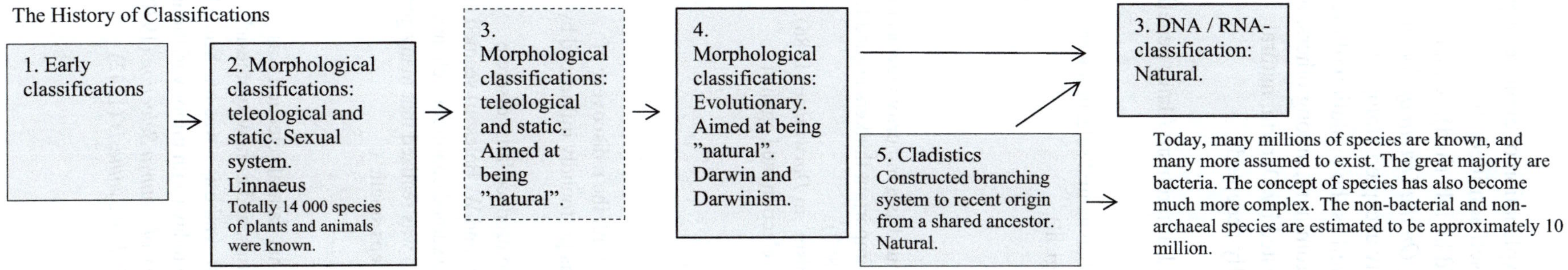

The Tree of Life

Pre-1800 there existed no tree of life (since one had a static view of biology) – those of Lamarck and Darwin's drawing in his notebook from 1837 are the first evolutionary versions).

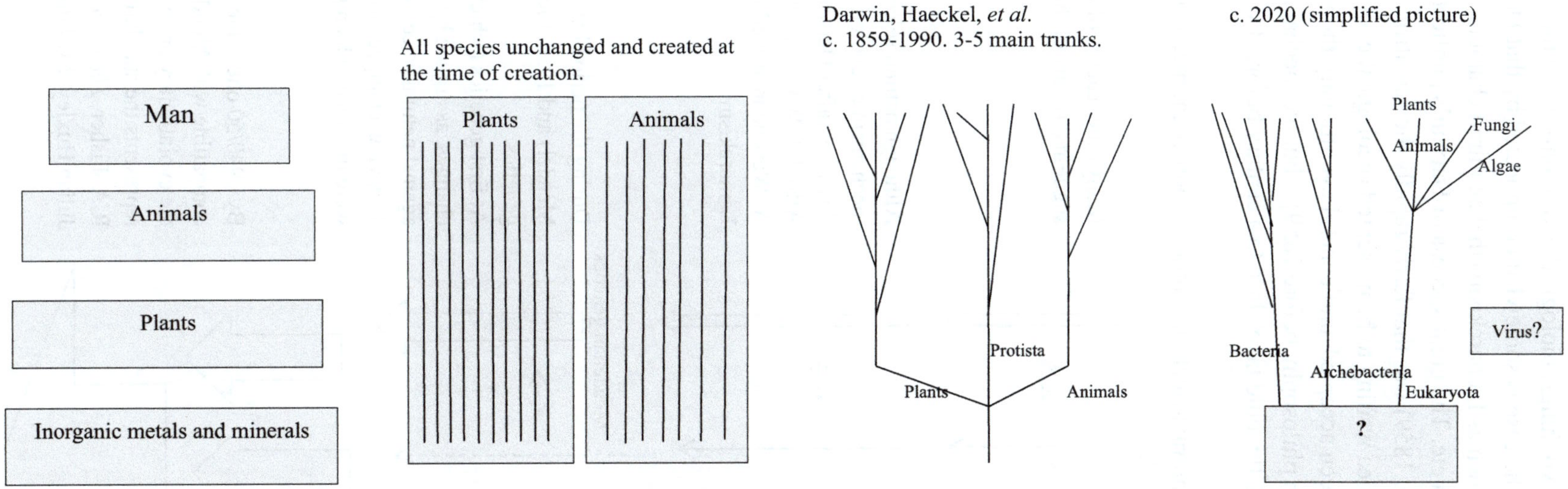

There is no clear consensus at present about the form of the tree of life. It is probably still a useful approximate for the eukaryotic (cells with nuclei) branch, but not for the others, where horizontal gene transfer occurs.

History of Darwinism

Darwinism usually means two things, biological evolutionism and the cause or explanation of evolutionary change, natural selection. The case is somewhat complicated by that Darwin himself admitted other mechanisms of evolution, such as Lamarckism (that acquired characteristics could be inherited), sexual selection, directed variation etc. The great success of Darwin's publication of the *On the Origin of Species by Means of Natural Selection*, 1859) was that fairly rapidly the scientific community accepted biological evolution – which was an idea waiting to be presented (geological evolution as well as the evolution of the solar system had recently been accepted, and general evolutionary thought was prominent, among others, in the writing of history and in philosophy before 1859) – but the mechanism suggested by Darwin, natural selection, was *not* accepted. It would take close to a century before it was generally accepted.

Darwinism = evolution + natural selection + other hypothetical mechanisms, e.g. Lamarchism / later genetics

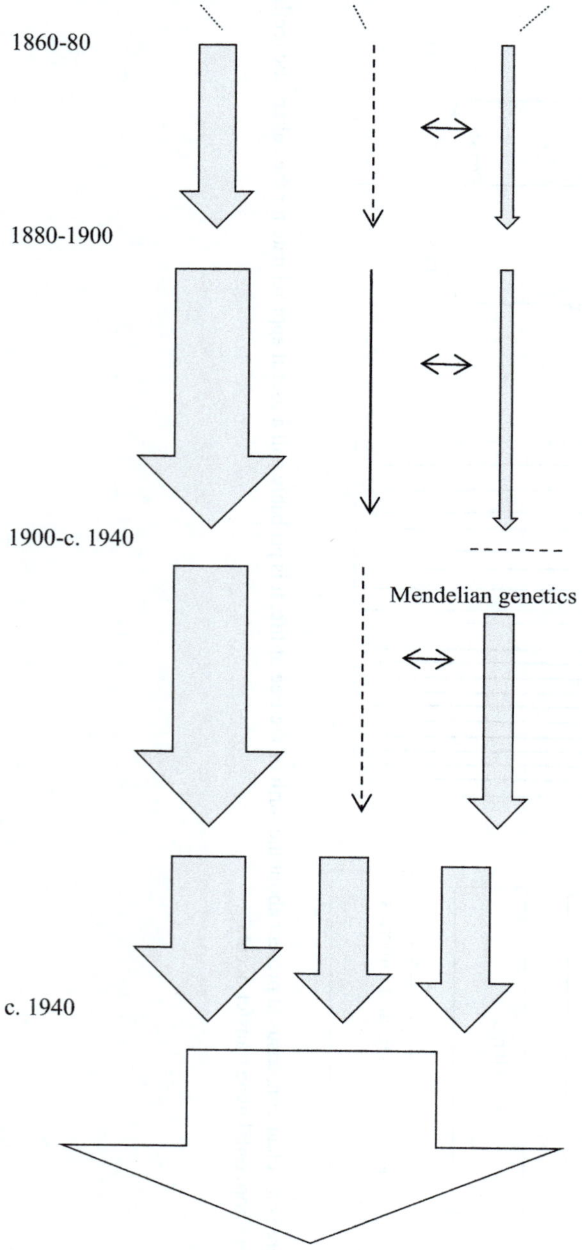

Many other causes of evolution than natural selection were suggested (many of them teleological).

Although natural selection continued as a minority view as the mechanism of evolution, several important thinkers argued for this view (after Darwin's death 1882):
Alfred R. Wallace, "pure Darwinism" in *Darwinism* (1886).
A. Weismann argued for natural selection, and against Lamarckism.

During the period between 1900 (with the rediscovery of Mendel) and 1930, the debate was profoundly influenced by genetics.
At first genetics (which seemed to produce large discrete changes, as argued by H. de Vries and W. Bateson) seemed to go against natural selection (which assumed continuous change). This view changed slowly when it was realized that many genes were involved, which made the steps small.

By 1940/50 one came to realize that natural selection was compatible with Mendelian genetics, and a "new synthesis" – the evolutionary or neo-Darwinist synthesis – was reached. This represents the standard view also today (with minor variations).
R.A. Fisher, *The Genetical Theory of Natural Selection* (1930)
Julian Huxley, *Evolution: The Modern Synthesis* (1942)

History of Evolutionism and Darwinism

Before c. 1750 there was very little awareness of biological fundamental change and evolution (although pre-Socratics and Epicurean Greeks suggested it). Most people during and after the Middle Ages assumed that the world was created about 4000 years B.C., and there was no awareness of a prehistoric era, all species were assumed to have been created at the same time, etc. Many believed that nature (and the universe) was created *for* man.

1750-1800. An increasingly dynamic conception of the universe, the earth and life began to develop. For example:

Astronomy / Cosmology: Buffon, Kant-Laplace nebular-hypothesis (however, the idea of the evolution of the whole cosmos, the Big Bang theory, had to wait until 1915-1970 to be suggested and accepted)

Geology: Buffon (age of the earth), Hutton (slow geological processes), and later, Hutton and Lyell

Biology: Buffon, Erasmus Darwin

Philosophy: Herder and Romantic *Naturphilosophie*, especially Goethe, and later, Hegel

1800-1820:

Jean Baptiste Lamarck (1744-1829) constructs the first synthetic theory of organic change, which he presented in *Zoological Philosophy* (1809). He also claimed that life begins spontaneously and that *species change*. Three factors important:

1. Life-force (vitalism)
2. Change according to the environment
3. Acquired properties are inherited.

1830-33. Charles Lyell (continuing the work of James Hutton), *Principles of Geology*. Implies that the earth's age is very great. Gave naturalistic explanations for earth's history.

1844. Robert Chamber's *Vestiges of the Natural History of Creation*, which speculatively in several ways foreshadows Darwin's work, and which had some similarity to German *Naturphilosophie.*

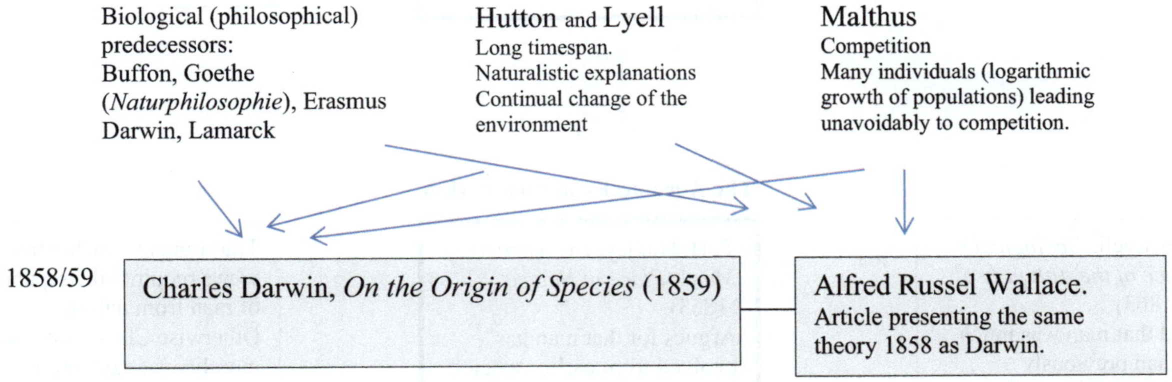

In the book, Darwin argues for that species change and develop. All species come from one or a few original ones. Gradual and progressive change, achieved through competition.
Assumes variation of offspring (empirically observed, but no satisfactory mechanism for this was given).

Survival of the fittest + sexual reproduction = Natural selection.

Darwin had avoided speaking about man in the *Origin of Species*, but did so in *The Descent of Man* (1871).

The Darwinian Revolution: Consequences in Three Fields

The consequences of the Darwinian revolution was enormous, not only for biology ("nothing in biology makes any sense without Darwinism"), but also for the general worldview and for our view of man's relationship to nature.

Biology

Lamarckism etc.
During the first half of the nineteenth century a certain awareness of the changes and transformations of species developed, especially Lamarckism. This either lacked or had a faulty explanation for why/how this change occurred or it was teleological (metaphysical, caused by the nature of reality)

Conflict with

Darwin, 1859
On the Origin of Species

1. Species change, evolve and develop. Evolution

2. This is caused by natural selection (competition)
a) Survival of the fittest
b) Sexual reproduction

Conflict with

Earlier and contemporary biology (and theology) which assumed that each species was created separately and did not change.

Worldview

Antropocentric worldview

The world changes, but the changes are for the better (for humans).
Man as separate from nature, at least the soul/consciousness is.
Non-materialistic.

Evolutionary naturalism

Darwinism (Huxley *et al*)

Materialism

A-teleological

Dynamic (everything changes)

"The modern worldview" (Spencer)

Christianity and the Church.

Many aspects of the Christian worldview was questioned:
Theological and teleological assumptions.
Traditional and Christian views of morality.
The world as created and basically static and stable.
Belief in 'eternal' truths and values (idealism).

The Age and Evolution of Man

Charles Lyell, *Geological Evidence of the Antiquity of Man* (1863)
Argued that man was much older than previously believed, but did not accept evolution and that man had evolved from earlier species.

T. H. Huxley, *Evidence as to Man's Place in Nature* (1863).
Argues for that man has evolved from earlier species, and for a new evolutionary ethics.

Traditional and Christian views required the separation of man from nature.
Otherwise Christianity and morality seemed impossible.
Man as separate from nature.
Man as created by God, in the image of God.

History of Physiology: 1. Overview

The Study of the Functions and Processes of Life

Aristotle
Galen

Jean Fernel, *On the Natural Parts of Medicine* (1542), defines physiology.

R. Colombo, 1559. Questioned Galen, especially the movement of blood (also Michael Servetus earlier).

W. Harvey, 1628. Dissections. Experiments. Circulation of the blood. This done in an Aristotelian philosophical framework. Harvey was a vitalist.

Many of Harvey's followers rejected Aristotelianism and teleology, and adopted mechanical philosophy. Iatro-mechanistics.

Borelli measures the force exerted by the heart (1678). Led many others to perform similar experiments, and to debate.

Vitalism, with teleology.

Physiology begins to become an independent discipline through Albrecht von Haller's textbooks on physiology 1747-76.

Between 1800-50 physiology becomes its own discipline, the first thoroughly experimental life science, with professors etc. Three different approaches led to this, primarily in France and Germany (Britain following later):

Naturphilosophie

Physiological Chemistry
New developments in chemistry made it possible to better understand physiological processes, e.g. respiration (Lavoisier and Laplace, 1780), digestion (1820s, Tiedemann and Gmelin) and K. Lehmann (1840s).

Quantitative Physiology
The use of quantitative experimental methods and new instruments, incl. electricity, developed physiology.
J. Müller, c. 1830. Influential textbook and many students:
Emil Du Bois-Raymond, c. 1840 (the physiology of electrical currents).
H. von Helmholtz, c. 1850 (the velocity of nerve impulses).
Carl Ludwig, c. 1850.
They wanted to reduce physiology to chemistry and physics; mechanists.

Vivisection
The extensive use of experiments on live animals greatly aided physiological knowledge.
L. Spallanzoni, c. 1770
(X. Bichat, c. 1800)
F. Magendie, c. 1820
C. Bernard, c. 1850

Anti-Vivisectionist Movement
Anti-vivisectionists (sometimes also anti-science) were active around 1850-75, but eventually they failed.

Many of the French physiologists, incl. Bernard, were vitalists (believing that living things possess a special vital force) or anti-reductionists.

During the second half of the 19th century physiology was highly successful (especially in Germany), and attracted many students (including foreigners) to its schools and laboratories, and spread experimental and quantitative physiology widely.

Around 1900 increased specialization led to that new fields separated off from physiology, such as biochemistry, pharmacology, cell biology, neurology, endocrinology and immunology, etc, but it also continued as a separate field, with influential researchers such as I. Pavlov, c. 1890 and Ch. Sherrington, c. 1900, and others.

History of Physiology: 2. Overview 1500–1900

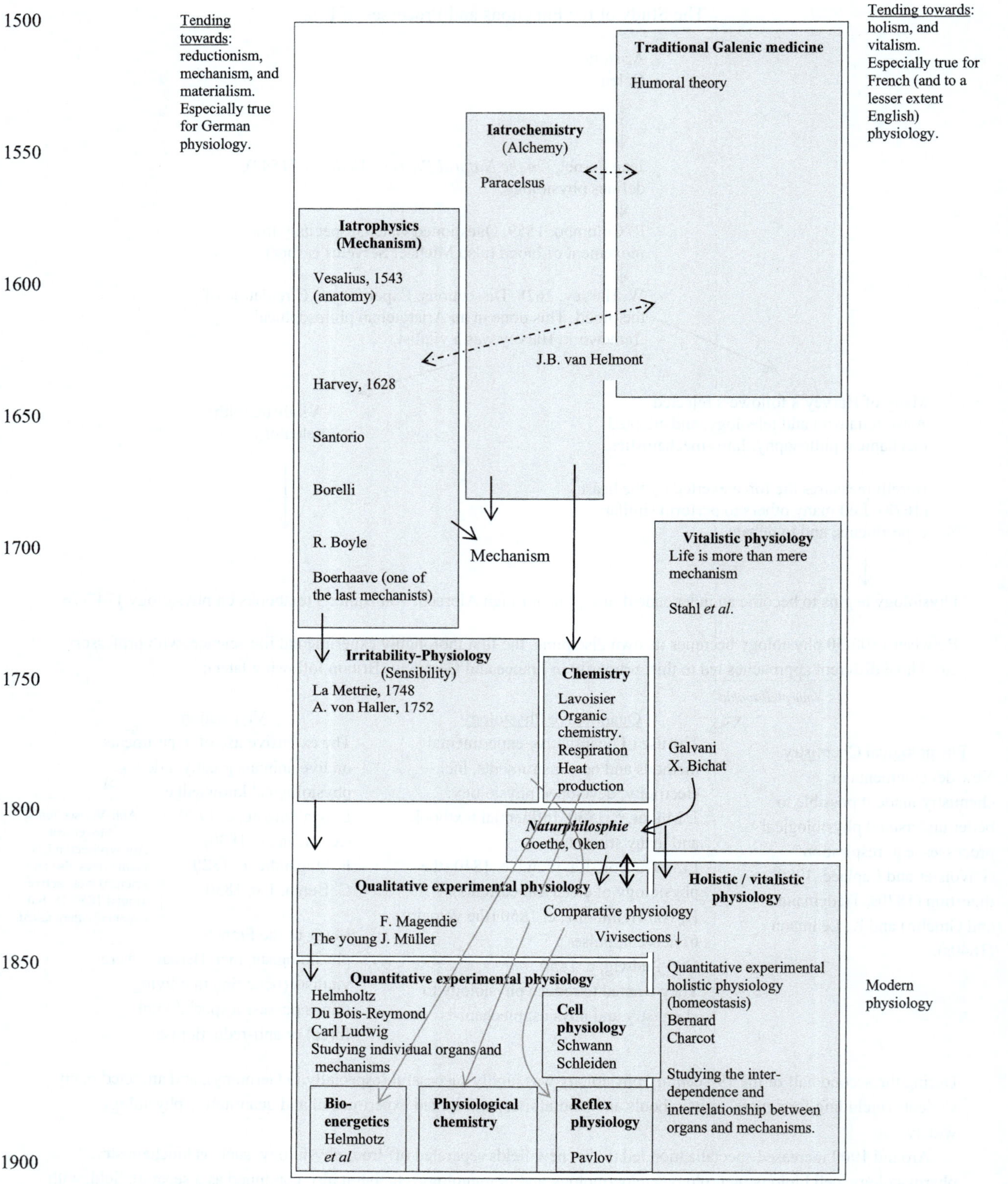

During the 20th century physiology has become less prominent since so many new fields have grown out of and separated from it, such as; cell biology, biochemistry, molecular biology, biophysics, endocrinology, neurology, etc.

NB. The space allotted to the 19th century has been enlarged to accommodate the text needed for this period.

History of Physiology: 3. Overview with Influences 1500–1900

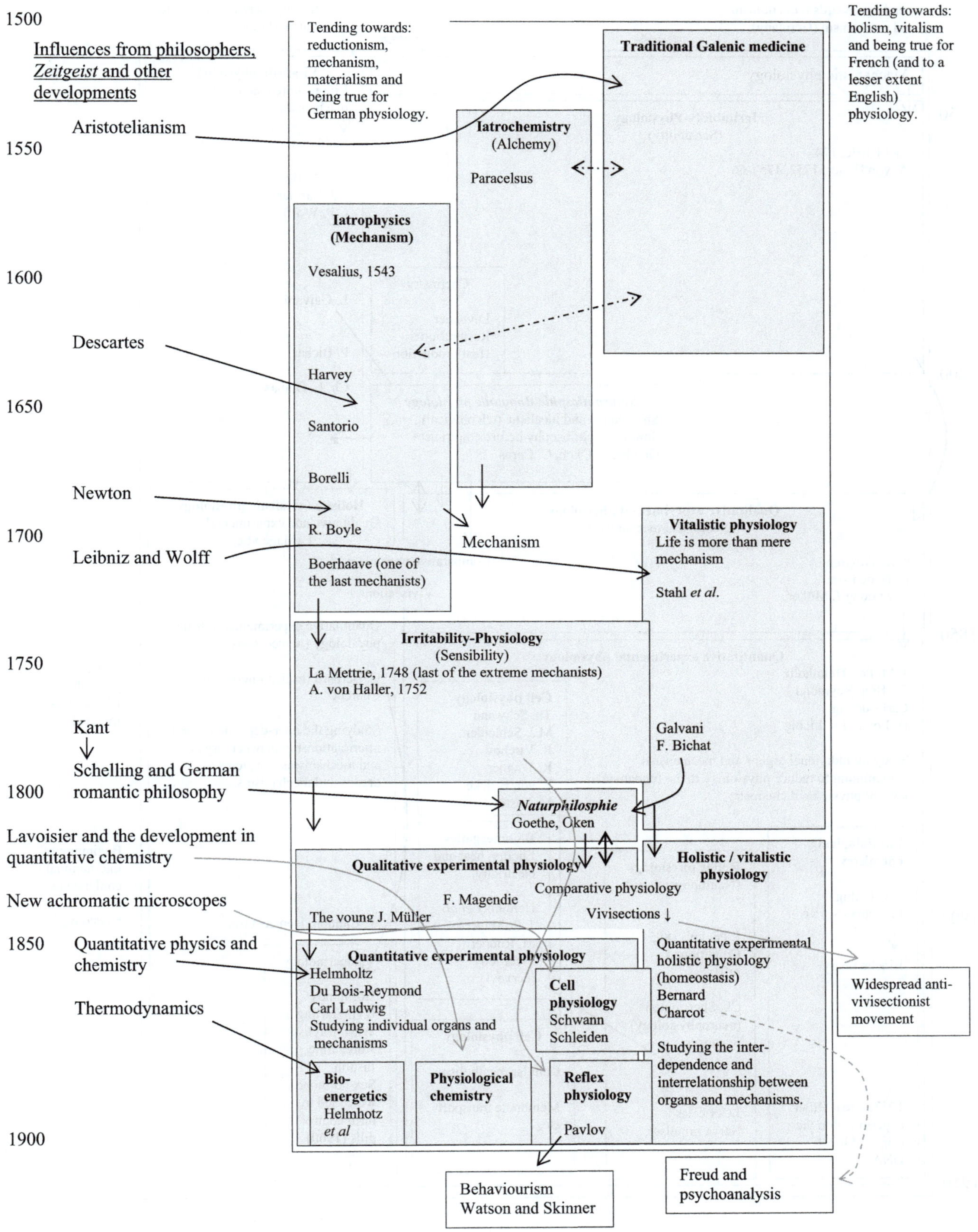

NB. The space allotted to the 19th century has been enlarged to accommodate the text needed for this period.

History of Physiology: 4. Overview 1750–1950

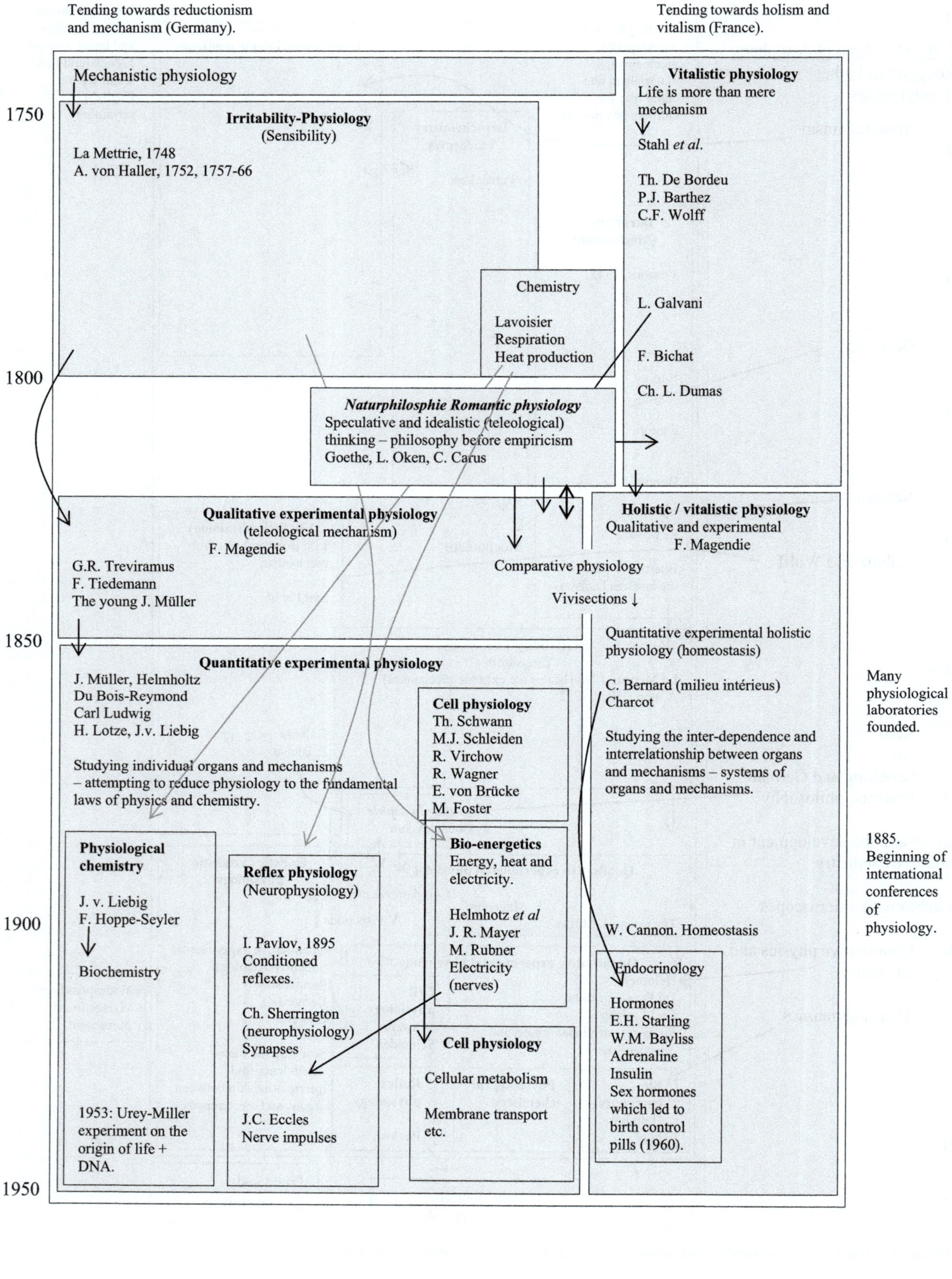

History of the Standard Scientific Theory of the Origin of Life

During the eighteenth and nineteenth centuries thinkers began to propose different (materialistic-mechanistic) theories that life could originally have arisen from non-life (already suggested by early Greek thinkers, including the Epicureans). However, much of these discussions were swamped by the polemical debates between materialists and vitalists about the spontaneous generation of life here and now. The problem of the origin of life is far from solved, but there exists a widely accepted general view of how it is likely to have occurred.

The Mechanism-Vitalism Debate
This debate colored much of the discussion also of the origin of life. Pasteur, 1862, generally regarded as having won for the mechanists.

Wöhler's synthesis of urea, in 1838, an argument for that life is not fundamentally different from the inorganic.

Darwin, 1859, 1871
Darwin avoided public discussion of the origin of life, but assumed that all life is related, and having one source. Darwinism played an important role in furthering and explaining how life develops mechanistically

Panspermia Theory
The panspermia theory, that life originated elsewhere in the universe was fairly popular. S. Arrhenius first to develop it scientifically, 1907.

The Primordial Soup Theory
In the 1920s A. Oparin and J.B.S. Haldane independently proposed that life could have originated in small ponds or in the sea

Alternatives:
Perhaps life instead originated among "dark smokers" (submarine hydrothermal vents) or among clays (combined with radioactivity).

1952/53 S. **Miller** (Urey-Miller) **experiments** which showed that complex organic molecules could (easily) be formed from simple starting molecules. Many organic molecules were found, including 5 different amino acids (later well over 25 amino acids found), the building material of proteins.

DNA, 1953
Until Watson and Crick discovered the structure of DNA in 1953 it was not understood how life functioned and reproduced.
1957. Crick. Three units for each amino acid.
The Central Dogma: DNA→RNA→Proteins
Suggested that probably RNA (or simpler related molecule) was the original form.
Hypothesis about necessary container, early cell structure.

The Origin of Life?
Work in progress, but no new fundamental step taken. Continued small progress in the understanding of many of the processes that probably are necessary for the origin of life.

History of Cell Biology

The cell constitutes the fundamental unit of life, and it plays a similar conceptual role in biology as elementary particles do in physics and atoms in chemistry. Cell biology began with work on the microscope in the 17th century. However, thereafter, during most of the 18th century, there was little progress, since most interest was focused on classification, until early in the 19th century due to improved microscopes and more organized and cooperative work.

R. Hooke, *Micrographia* (1665). Coins the name cell from his study of cork with his microscope.

↓

A.v. Leeuwenhoek, 1673. Discovered spermatozoa, bacteria, sperm cell, etc. Used a small single lens microscope.

↓

Early in the 19th century *Naturphilosophen* (including Oken) argued that all biological material consists of *mucus vesicles*.

↓

R. Brown found the cell nucleus in plant cells in 1831, but neither he nor others made much of this at the time.

In the 1820s the improved achromatic microscope was developed, which led to a renewal and expansion of cell studies.

↓

Mathias Schleiden, 1838, proposed that the cell was the basic unit of all plants.

↓

Theodore Schwann, 1839, claimed that the cell also constituted the unit of all animals.

↓

All life thus based on, consists of, cells. Cells were regarded as autonomous living entities. All organisms started life as a single cell and develop by the formation of new cells. Both Schleiden and Schwann assumed that cell formation occurred outside the cell.

↓

1840s. Many researchers: von Mohl, Nägeli, Remak, *et al*. New cells not formed outside or inside old cells (as previously believed) but by fission, by cell division (mitosis).

→ The bacterial thesis of disease. Pasteur and Koch, c. 1870s-80s.

↓

Led to the general development of vaccination (inoculation). (This had already been performed a few times during the 18th century.)

R. Virchow popularized cell theory: *Omnis cellula e cellula* – all cells from cells. Virchow used cell theory in medical pathology and histology, and suggested that disease began in one individual cell. An organ and an organism consist of the unity of many autonomous cells. Virchow never accepted the importance of bacteria.

↓

(The chart continues on the next page)

1870s and 1880s. Chromosomes discovered, and a better understanding of the role of the nucleus in normal cell division (mitosis).

→ Other Cell Organelles discovered
Mitochondria, ribosomes, etc

↓

A. Weismann, c. 1890, and earlier H. de Vries, assumed that the nucleus /chromosomes constitute hereditary particles.

Mendel's genetic work rediscovered, 1900.

↓

c. 1910. T. H. Morgan *et al* study the relation between genes and chromosomes in fruitflies. First chromosome map 1912. The gene no longer an abstract unit, but a segment of the chromosome.

↓

One gene – one enzyme/protein hypothesis

↓

The nucleus consists of an unusual and large molecule.

↓

The discovery of DNA, 1953

↓

Genetics
Chemical and biological understanding of genetics.

History of Genetics: 1. Different Hypothesis and Theories

Until at least c. 1900 there was uncertainty and confusion as to how heredity worked, including what "units" were actually inherited.

Darwin, *On the Origin of Species*, 1859

Doubs and discussions about Darwin's theory made the question of the mechanisms of inheritance acute, 1859-1940.

Mendel, 1865/66
Discovers the Mendelian laws of inheritance and determined seven units which were inherited independently of each other. But received no attention.

F. Galton, c. 1880
Biometrics and statistical analysis.
Weldon and Pearson, 1900

Five theories of inheritance, 1800-1900:

1. Blending Theory
The hereditary material of the parents averages or blends in the offspring. E.g. Darwin, Nägeli.

2. Pangenesis
The inheritance of characteristics (incl. those acquired) by particles in the cells. E.g. Darwin

3. Preformation
God or nature has determined the shape of all offspring already at the creation of the world (the form of the offspring is already present in the germ).

4. Mechanism
The shape of the offspring determined by the laws of physics and chemistry, but not yet understood.

5. Vitalism
A vital force determines the organization of organic matter. Including epigenesists.

A. Weismann. 1880s and 90s. Rejected inheritance of acquired characteristics – and accepted sole role for natural selection. Strict version of Darwinism. Cell nucleus and its chromosomes as the site of inheritance, and they not affected by changes in life style.

1900. Mendelian laws rediscovered by de Vries and C. Correns.

The term "gene" is coined 1909 by W.L. Johannsen

Four different schools of how inheritance works, 1900-40:

1. Darwinism and NeoDarwinism
Natural selection main principle. Claimed that it works continuously .

2. Mendelianism
W. Bateson and R.C. Punnett names the field "genetics" in 1906. Inheritance is done in small discrete steps.

3. Mutationism.
Discrete evolution in large radical steps argued by Hugo de Vries. Faster evolution.

4. Lamarckism and Neo-Lamarckism.
Inheritance of acquired characteristics.
Orthogenesis, i.e. that evolution is directed by inner force (e.g. to greater complexity)

1910-15. Th. H. Morgan *et al.* Work on fruit flies. Argues for "the Mendelian-chromosome theory" of heredity. Genetic steps are small. Genetic maps of chromosomes. *The Mechanism of Mendelian Heredity* (1915).

1930-40/50. "The new synthesis", the neo-Darwinian and Mendelian synthesis.

O.T. Avery *et al*, as part of the Delbrück phage group, showed in 1944 that nucleic acid carries the genetic information – previously assumed to be done by proteins.

G. Beadle, 1945, One gene, one enzyme-hypothesis.

Linus Pauling

1953. DNA. Double helix structure. R. Franklin. Watson and Crick.

1960-65. Discovery of RNA, mRNA and tRNA.

1964. The genetic code: three base pairs in the DNA and RNA code for specific amino acids.

Increased complexity of gene activity (knowing the genetic code not enough) – can be context-dependent etc.

Molecular biology.
(See chart below)

Determining full genomes.
HUGO (2001) and others.

History of Genetics: 2. Discoveries

Until at least c. 1900 there was uncertainty and confusion as to how heredity works, including which "units" are actually inherited.

Mendel, 1865/66
Discovers the Mendelian laws of inheritance and determined seven units which were inherited independently of each other. But received no attention.

Statistical analysis

O. Hertwig 1875 showed that the embryo grows from a single cell from the female fertilized by a single male sperm.

F. Galton, c. 1880
Biometrics and statistical analysis.

E. van Beneden showed in 1882 that egg and sperm cells received a single strand of the normally paired chromosomes (meiosis).

W. Flemming. Observed chromosomes in normal cell division (mitosis) in 1882 for the first time.

A. Weismann. 1880s and 90s. Cell nucleus and its chromosomes as the site of inheritance, and that they cannot be affected by changes in life style.

1905. Nettie Stevens. The sex of the offspring determined by the chromosomes carried by the sperm.

1901. Mendelian laws rediscovered by de Vries and C. Correns.

Weldon and Pearson, 1900

1908. H. Nilsson-Ehle shows that properties that depend on two or three genes follow Mendelian laws, but the steps become much smaller.

1910-15. Th. H. Morgan *et al.* Work on fruit flies. Argues for "the Mendelian-chromosome theory" of heredity. Genetic steps are small. Genetic maps of chromosomes. *The Mechanism of Mendelian Heredity* (1915).

Fisher, 1915, 1930

1919. W. Castle showed that slight variations in rat coating was genetically stable – did not return to "average". Castle became a champion of natural selection.

O.T. Avery *et al*, as part of the Delbrück phage group, showed in 1944 that nucleic acid carries the genetic information – previously assumed to be done by proteins.

G. Beadle, 1945, One gene, one enzyme-hypothesis.

Linus Pauling <->

1953. DNA. Double helix structure. R. Franklin. Watson and Crick.

1961. F. Jacob and J. Monod showed that some genes can regulate the expression of other genes. Hypothesized the existence of mRNA, found soon afterwards.

1960-65. Discovery of RNA, mRNA and tRNA.

1964. The genetic code: three base pairs in the DNA and RNA code for specific amino acids.

Increased complexity of gene activity (knowing the genetic code not enough) – can be context-dependent etc.

Genome projects, inclusive HUGO (2001).

The Central Dogma of Modern Genetics: 1. Overview

The central dogma of molecular biology is the explanation of the flow of genetic sequence information within all living biological systems. With words, it is summarized as: DNA makes RNA, and RNA makes proteins. This was first formulated by Francis Crick in 1957.

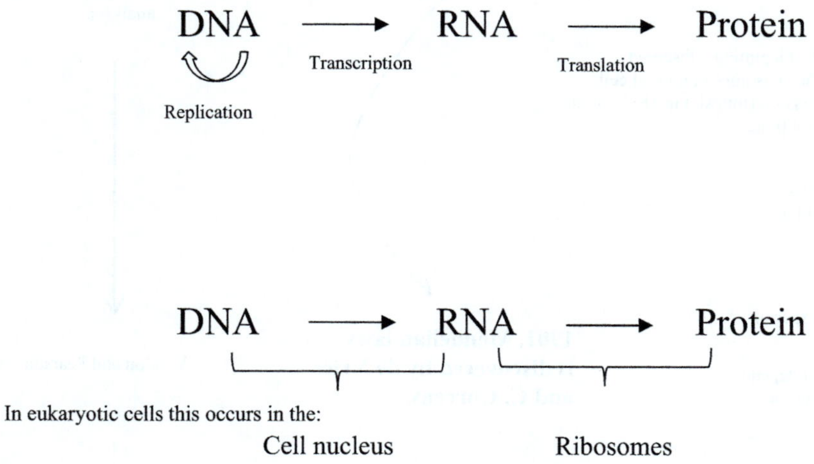

Proteins are large biomolecules that consist of one or more long chains of normally twenty different kinds of amino acid residues. They are the most abundant and functionally diverse molecules in living systems. Almost every life process depends on this class of molecules. For example, enzymes and polypeptide hormones direct and regulate metabolism in the body. Contractile proteins in muscle enable movement. The protein collagen forms in bone a framework for the deposition of calcium phosphate crystals. In the bloodstream, proteins transport molecules essential for life.

History of Ecology: 1. Outline

E. Haeckel, the German Darwinist, coined the word 'ecology' in 1866 for 'scientific natural history', but it really only began as a field of study in the 1890s.

Influences and predecessors

Darwin, 1859
Evolution

Alexander von Humboldt, c. 1800
Plant geography (the importance of geography)
All of nature interconnected.

↘ ↙

Ecology

From the very beginning there were strong tensions in the field:

Darwinism, mechanism, materialism <-·-·-> Romantic holism

Laboratory work
Physiology, genetics <-·-·-> Field work
Taxonomy, natural history-tradition

The earliest influential ecological works were textbooks in plant ecology by Eugenius Warming, 1895 and by A.F.W. Schimper, 1898. The treatment of animals came later, and they began to be treated together in the 1930s, when also the term 'ecosystem' came into use.

In several European countries a 'plant sociology' approach were used, including by Warming, Forbes and Cowles, which was more pluralistic and often did not assume a final stable state.
Tansley, ecosystem, 1935.
Elton.
J. Huxley.

F. E. Clements (1905). Argued for nature as a 'climax' community. Describing the dynamic approach until a climatically determined 'climax' community was reached, as an endpoint, as the 'natural' and static "final" state (stable, self-sustaining, equilibrium state).

<u>Alternative, somewhat "romantic", views:</u>
John Phillip's *Superorganism.*
J. Chr. Smuts, *Holism and Evolution* (1926).
W.C. Allee. Animals cooperate instead of compete.

Post-world war II:

Scientific ecology became significantly more Darwinian oriented.

The use of radioactive isotopes led to much better energy and nourishment studies along the food chain.

Mathematical modeling became common (and critique of its assumptions).

After the 1960s (and especially post- Rachel Carson's *Silent Spring* (1962) there emerged a strong environmental movement which often attempted to base its results on ecology.

Energetics

Arthur G. Tansley, 1935
↓
R. Lindeman, 1942. Ecosystem synthesis (with no 'climax')
↓
Howard and Eugene Odum, *Fundamentals of Ecology* (1953)
↓

History of Ecology: 2. 1890–2000

Two Traditions Between Two Philosophical Positions

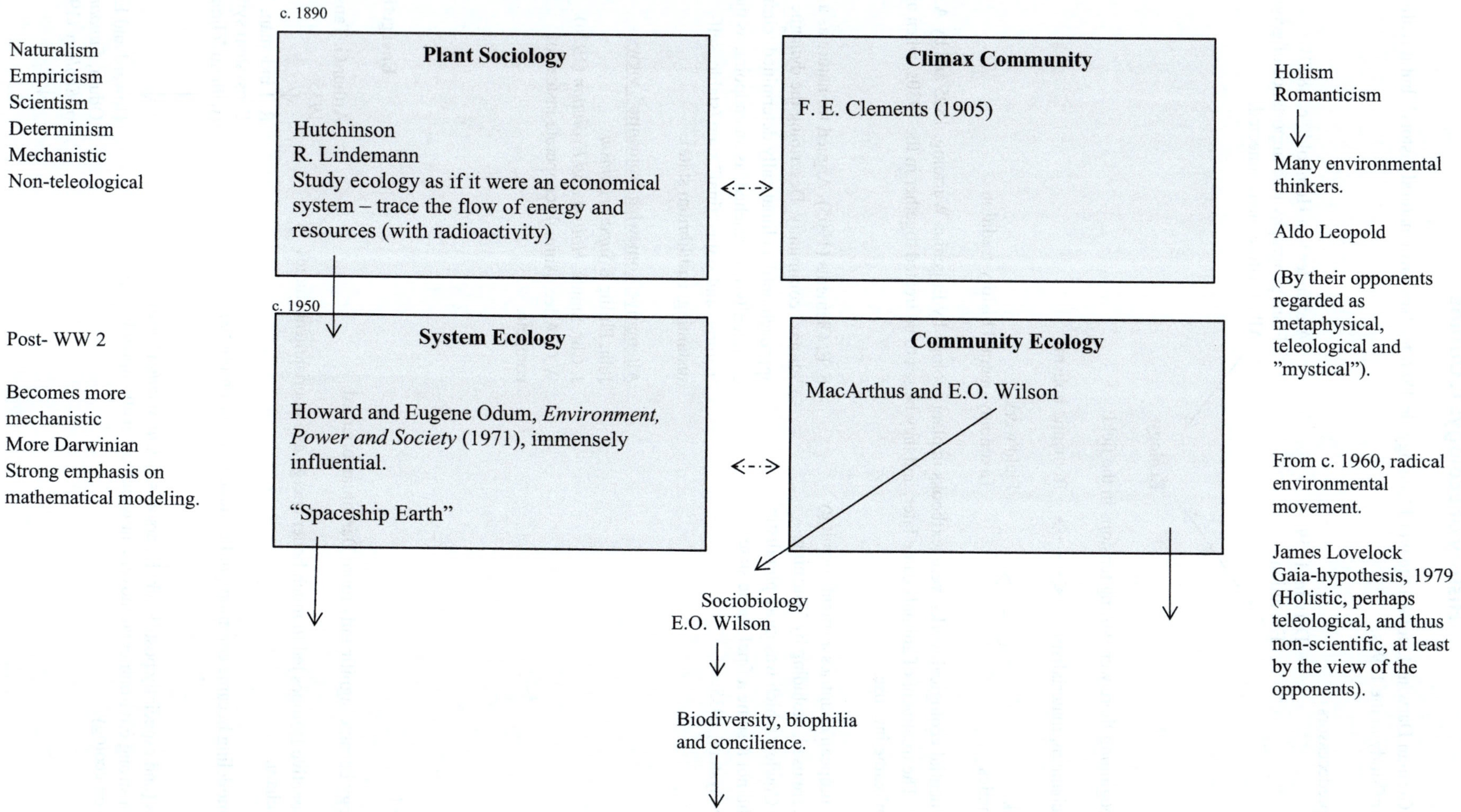

History of Entomology (Knowledge of Insects): 1. Outline

Not Chronologically Proportional

Aristotle, c. 340 BC
The father of entomology
He described some sixty *entoma*, insects. Attempted a natural system of insects, which lasted for over 2 000 years.
Made many correct (and some incorrect, such as spontaneous generation) observations of insects in his *Historia animalium*.

↓

Pliny the Elder, *Historia naturalis* (AD 79)

↓

Albertus Magnus, *De animalibus* (1250)

↓

Systematic Entomology (taxonomy) | Microscopic studies of insects (anatomy)

c. 1650. Experimental method and the microscope

Systematic Entomology (taxonomy)

John Ray, c. 1660-90
Some insect systematics.

C. Linnaeus. Founder of systematic entomology, c. 1735-58-70, using wings (-ptera).
First four, later seven, orders of insects (today 29 are generally recognized).
Linnaeus' system influential, but insects not his main interest.

Many new more detailed systematic treatments and ways to systematize, c. 1770-1850 (including by Linnaeus' students and followers).

P.A. Latreille, c. 1810, divided the insects (bugs) into three classes: crustacea, arachnida and insecta (in 8 orders).

J.T. Lacordaire unified the classification of coleuptera (beetles, the largest group).

Microscopic studies of insects (anatomy)

Many microscopic studies of insects

J. Swammerdam; A.v. Leeuwenhoek; R. Hooke; F. Redi
c. 1670 ; c. 1680-1720 ; 1665 ; 1668

Further increasingly detailed microscopic studies during the 18th century:
R.v. Rosenhof; P. Lyonet; Réaumur; C.K. Sprengel (c. 1793, discovered the natural dependence of insects and flowers); J. Lamarck; G. Cuvier.

Microscopes improved c. 1810.

Collections of insects used in Academies and helped create early museums.

Several even more detailed studies of insect anatomy and phylogeny: e.g. H.E. Straus-Duckheim, L. Dufour, W. Kirby and W. Spence, *Introduction to Entomology* (1816-26).

Darwin, 1859. Used his study of insects to support his theory.

Insect mimicry gave strong support for evolution in 1860-80.

History of Entomology: 2. Overview

Not Chronologically Proportional before 1600

Work Mostly by Naturalists

Aristotle

↓

Pliny

↓

1600

1700

R.A. de Réaumur (1683-1757). Produced a detailed study, with drawings (6 vols., 1734-42).

Carl Linnaeus *Systema Naturae* (1735). Insect classification by using wings (-ptera) and binomial nomenclature. First four classes, later seven.

P.A. Latreille (1762-1833).

G. Cuvier (1769-1832).

J-P Lamarck (1744-1832). Several works. 1801 and 1810.

G. White, *Natural History of Selborne* (1789).

1800

W. Kirby and W. Spence, *Introduction to Entomology* (1816-26).

Ch. Darwin, *Origin of Species* (1859). Some of Darwin's arguments based on insects.

↓

Work Mostly by Microscopists

Robert Hooke, *Micrographia* (1665).

A. van Leeuwenhock (1632-1723). Microscopic studies of insect morphology and insect organs.

1668. F. Redi. Maggots come from eggs of flies, not spontaneous generation.

Jan Swammerdam (1637-1680). Anatomical studies of insects, especially the honey bee.

↓

Continued work with microscope, which eventually merges with that of the 'pure' naturalists.

↙ ↓

The Debate Between Materialism and Vitalism

Materialists typically regard everything, including life and the soul (if it exists) as matter in motion. Vitalists argue that life is special, being dependent on an immaterial soul or an irreducible life force.
The debate between materialists and vitalists was especially intensive between c. 1750-1850, and was eventually won by the materialists and positivists who argued that the hypothesis of an immaterial soul or life-force was either wrong or superfluous.
Both sides of the debate go back to antiquity. The term 'vitalism' first occurred in the 17th century in reference to older theories which posited life-bearing seeds or spirits, such as those by Paraculsus, Robert Fludd and J.B. van Helmont.

Materialists		**Vitalists**
Democritus Epicurus Lucretius		Aristotle Galen
[c. 200-c. 1600. There were essentially no materialists during this long period.] Revived materialism: Hobbes Gassendi (man has a soul, but not other lower forms of life)	<-·-·>	Vitalism, or kindered views, was the standard in both Aristotelianism and Christianity, i.e. until at least c. 1600. The term 'vitalism' begins to be used around 1650. Most older views, those of Plato, Aristotle, etc, were interpreted as vitalistic.
Mechanists		Harvey, 1628
The Enlightenment: Many materialists, e.g. La Mettrie and d'Holbach. During the 19th century there were an increasing number who criticized vitalism: Pasteur Helmholtz Büchner Moleschott Darwin Huxley	<-·-·> <-·-·>	Following Georg Ernst Stahl, c. 1700, the German physician and chemist (who also introduced phlogiston) vitalism became a common theory among physicians during the 1700s. It was also often supported by many romantics (as well as many Christians and generally conservatives). With the decline of romanticism, c. 1830s, secularization, and advances in medicine and physiology there were fewer and fewer avowed vitalists after c. 1850.
		Buffon, 1750 Caspar Wolff, 1775 J.F. Blumenbach, 1800 L. Oken, 1820 K.E. von Baer, 1850 Hans Driesch, 1920 These thinkers continued to argue for vitalism, although it had by now become a distinctly minority view among natural scientists.

Views of the Age and of the History of the Earth (Geology): 1. Overview

Before the 18th century there existed two main and highly contradictory views of the age of the earth:

Ancient Greek view:
The earth is very old or even eternal.

A view held by e.g. the pre-Socratics, Aristotle, Epicurus, Ovid, Lucretius (and which influenced e.g. G. Bruno, executed 1600).

From the Middle Ages and onward the age of the earth was calculated according to the Biblical account which gave a very short history of the earth.

The paragon here is James Ussher, who, c. 1650, calculated that the earth was created 4004 B.C.

This view became the generally accepted view in Europe.

Buffon (1707-1788), in his *Historie naturelle* (1749) suggests that the earth was formed from material from the sun after a comet had hit it, and that it would require about 75 000 years to cool (the beginning of an evolutionary theory of geology). In 1778 he suggests that the earth is perhaps several million years old.

Two major themes of debate concerning the forces and events that had shaped the earth occupied much time and space during the 18th and 19th centuries:

Plutonism / Vulcanism ⟷ Neptunism
("Dry" processes on land) (Water, esp. great floods)

Uniformitarians ⟷ Catastrophism
(Slow processes, still active) (Catastrophes most important)

James Hutton (1726-1797)
Theory of the Earth (1795) Slow processes.
Plutonism

⟷ Abraham Werner (1750-1817)
Catastrophism and Neptunism.

John Playfair
Popularises Hutton, 1802

Charles Lyell (1797-1875)
Principles of Geology (1830). Continual slow processes. Study processes which are active now.

↔ G. Cuvier, 1812
Catastrophism and Neptunism.
Paleontologist. Studied fossils of extinct species. Use of fossils to date rocks.

R. Jameson. Religious geologist.
W. Buckland. Found fossils of gigantic lizards (dinosaurs). R. Owen coined the term 1842.

Darwin, 1859
Darwinism required long time-periods to work as an explanation.

↕

Problems with the age of the earth. c. 1870-1905. Lord Kelvin. A cooling star. 20-40 million years old (not enough for Darwinian evolution). Changed to 100. Mountains arise, and the surface of the earth shaped, when the earth cools.

⟷ Plate-tectonics: Explains how mountain-ranges are formed and how the earth surface is shaped.
Alfred Wegener, *The Origin of Continents and Oceans*, 1915.
Not accepted. No explanation.
1927. A. Holmes. Convection.
Plate-tectonics accepted during the 1960s.

Radioactive dating of earth
1896. Becquerel. Radioactivity

1905-07. Radioactive dating.
Rutherford *et al*, 100-600 mil.
1921-27. Holmes. 1.6-3 billion
1956. 4.5 billion years old.

Views of the Age and of the History of the Earth (Geology): 2. 1700–2000

The hub of geology in the 18th and 19th centuries was stratigraphy, the ordering of the different layers of rock. This was greatly aided by paleontology at the beginning of the 19th century, when fossils were used to date and order rock strata.

Two major themes of debate concerning the forces and events that had shaped the earth occupied much time and space during the 18th and 19th centuries:

Plutonism / Vulcanism ⟷ Neptunism
("Dry" processes on land) (Water, esp. great floods)
A. Moro, 1740. Woodward (religious), 1695
Debate especially 1790-1830

Uniformitarian ⟷ Catastrophist
(Slow processes, still active, Most important) (Catastrophes most important) (i.e. floods and/or volcanism)

Religious conceptions of the earth (such as Noah's Flood) and its age, c. 6000 years old.

James Hutton (1726-1797)
Theory of the Earth (1795) Slow processes. Plutonism. The earth as immensely old. He was criticized as being against the Bible and had relatively few followers 1800-1830.

Much field work was done 1790-1830, including using fossils to determine stratification chronology.

John Playfair
Popularises Hutton, 1802

James Hall, performs experiments which supported Hutton.

Charles Lyell (1797-1875)
Principles of Geology (1830-3). Continual slow processes. Study processes which are active now.

Abraham Werner (1750-1817)
Catastrophism and Neptunism. Stratigraphy. Many students and his theories highly influential during the beginning of the century.

G. Cuvier, 1812
Catastrophism and Neptunism. Stratigraphy
Paleontologist. Studied fossils of extinct species. Use of fossils to date rocks. *Essay on the Mineralogical Geography of Paris* (1810). Cuvier was highly influential and opposed both geological and biological evolutionism.

R. Jameson. Religious geologist.
W. Buckland. Found fossils of gigantic lizards (dinosaurs). R. Owen coined the term 1842.

Problems with the age of the earth. c. 1870-1905. Lord Kelvin. A cooling star. 20-40 million years old (not enough for Darwinian evolution). Changed to 100. Mountains arise, and the surface of the earth shaped, when the earth cools.

Plate-tectonics: Explains how mountain-ranges are formed and how the earth surface is shaped.
Alfred Wegener, *The Origin of Continents and Oceans*, 1915.
Not accepted. No explanation.
1927. A. Holmes. Convection.
Plate-tectonics accepted during the 1960s.

Radioactive dating of earth
1896. Becquerel. Radioactivity

1905-07. Radioactive dating. Rutherford *et al*, 100-600 mil.
1921-27. Holmes. 1.6-3 billion
1956. 4.5 billion years old.

Geophysics and Plate Tectonics

How Are Mountains Formed and Do Continents Move?

The standard view during the nineteenth century (and earlier) was that the earth as a totality is static (the continents have always been where they are now) and that mountains have formed due to the earth cooling and shrinking. E. Suess, in the 1880s, worked out and summarized a detailed picture of this, including with much paleontological evidence.

This was generally accepted, but by 1900 it was realized that it also contained problems, such as why mountain ranges mostly are at the edges of continents and with the discovery of radioactivity, it became uncertain if the earth really was cooling.

This box is large because of the amount of text. It really should be much smaller than the other box.

The Standard Static View

The view described above continued to be the standard view also during the period 1900-1960, but often in a less dogmatic manner than earlier.

↔

Continental Drift

Several speculative hypotheses of continental drift had been suggested before A. Wegener's in 1912, but his was the first more thorough and the first to receive considerable attention. He developed it further in *The Origin of Continents and Oceans* (1915, and several revised editions until his early death on Greenland, 1930).

Wegener could explain several problems with the standard view, and also the shapes of the continents, the similarities of living and fossil species, and between the geology, on both sides of separated continents, why mountains mostly are found along continental edges etc.

However, he admitted that he had no good explanation of the forces which caused the continents to move.

This hypothesis led to intensive debates and discussions. Some geologists supported Wegener, but the majority remained statists.

Among the statists, perhaps the most influential was H. Jeffreys, but also essentially all North American geologists belonged to this camp.

↔

Several geologists continued to support Wegener's thesis also after his death. Most important of these were A. Holmes (who began as believing in a static earth), who also suggested large-scale convection as solution to the "mechanism" problem (and constituting the origin to the "seafloor spreading" hypothesis). His *Principles of Physical Geology* (1944) ended with a chapter on continental drift.

During the 1950s there was standstill in the debate. However, important developments were made in paleomagnetism (the study of the history of changes in the earth magnetic field) and in oceanography (which made great advances due to its military importance). Oceanic ridges were found, and the main problem became to explain their origin. Many hypotheses were suggested, including "seafloor spreading" which was consistent with continental drift. From this hypothesis followed consequences concerning seafloor magnetism and the existence of "transfer faults" that could be tested. When both were confirmed in 1966, continental drift was accepted almost immediately.

The following year, 1967, plate tectonics were suggested by Jason Morgan and Dan McKenzie, independently, when seafloor spreading was applied to a spherical surface and it was realized that Euler's theorem could be used to describe the relative movements of circa ten rigid plates that comprise the earth's outer layer. Shortly afterwards seismological evidence for rigid plates of about 100 km thickness was found.

Continental Drift and Plate Tectonics 1860–1980

Increased Dynamism and Evolutionism in Geology

The static ("fixist") view The earth as shrinking		**Continental drift** "mobilists"
For example, E. Suess summarizes this view, the standard view, c. 1880s.	⟷	Early speculative theories of continental drift.
Mountains caused by the cooling earth shrinking.		A. Wegener , 1912, 1915-30. Complex theory of continental drift, but lacked explanation of how continents could move.
The discovery of radioactivity at the very end of the 19th century led to better knowledge about the age of the earth, early in the 20th century (A. Holmes *et al*) as being much older than assumed. It also led to questioning if the earth really is cooling (or, if it is, then it must be much slower than previously assumed). The great majority of geologists (especially American) continued to be opposed to the theory of continental drift, including H. Jeffreys and Ch. Schuchert.	⟷	⇩ Alex du Toit, *Our Wandering Continents* 1937 A. Holmes, 1928, 1944 (and a few other Europeans) supported continental drift, and suggested large-scale convection as possible source to movements.

During the 1950s one discovered extensive ocean ridges, and many possible explanations for them were suggested.

H. Hers and R. Dietz, themselves not supporters of continental drift, independently, gave several hypothetical explanations to ocean ridges → One of these, 1960-62, "seafloor speading", consistent with continental drift, where the continents do not plow through the seafloor, but simply follow its movement.

The hypothesis of "seafloor spreading" could be tested. 1. Seafloor paleomagnetic material symmetrical on both sides of the mid-ocean ridges. 2. Transform faults. Both confirmed in 1966.

→ This led to rapid acceptance of continental drift in the form of seafloor spreading.

↓

Already in 1967 seafloor spreading was combined by J. Morgan and D. McKenzie with the theory of movements of (7-8 major and many minor) solid plates (moving at a rate of few centimeter per year) yielding plate tectonics.

↓

The acceptance of plate tectonics came to revolutionize many aspects of the earth sciences.

The Discovery of Ice-Ages and Glaciation 1: Overview 1700–2000

The earth was believed to be relatively young and was assumed to be cooling – i.e. had been hotter earlier, which made ice ages unlikely.
The assumed great flood or floods was used to explain phenomena.

Jean-Pierre Perraudin, 1815, argues for more extensive glaciers.

↓

Jean de Charpendier further developed the theory and presented it during a conference 1834 (and in a book 1841)

↓

Louis Agassiz, paleontologist, expert on fish fossils, began as skeptic but was converted by Charpendier, and presented his views and studies during a conference 1837 and in the book *Études sur les glacier* (1840), where he suggested that there has been a great ice age.

↕ ↓

There was much resistance to this idea by geologists and others, but by 1860 the majority, including Ch. Lyell, had accepted it due to its ability to explain many geological phenomena.

However, one had no obvious cause for how and why increased glaciation and ice-ages occurred. → James Croll, *Climate and Time* (1875), argues for that it is changes in the shape of the earth's orbit which caused the ice age(s). This would imply that glaciation in the opposite sense on the southern hemisphere. This hypothesis accepted until c. 1920s.

Nonetheless, the ice age and ice bridges were used to explain many problems of geology and geography, including how people came to the Americas during prehistoric times.

↓

1920s. Agassiz's one ice age exchanged for four periods of glaciation – in both northern and southern hemisphere.
A succession of ice ages now generally accepted. → M. Milankovich calculated the amount of radiation received on earth per year during the past half million years.
Argued for radiation cycles during the 1920s, but not accepted then, but has been accepted later.

↓ ↕

Post-World War II.
Rejection of the four stage model of the 1920s as inconsistent with deep sea cores (where yearly sediments can be studied). Instead a realization that there have been a large number of ice ages (perhaps also one or several periods when the whole earth has been frozen, "snowball earth"), and that we live in one now, which began c. 2.6 million years ago. Within each ice age there are colder (glacials) and warmer (interglacials) periods. We are now in an interglacial period, with the last glacial maximum c. 20 000 years ago. Today there is no consensus of any simple cause for the major ice ages, as well as for the variations within these (between glacials and interglacials), but general agreement that they are due to many factors such as changes in the earth's orbit (Milankovitch cycles), atmospheric composition, the motion of tectonic plates (resulting in changes in wind and ocean currents etc), the orbital dynamics of the earth-moon system, the impact of large meteorites and the effect of vulcanism and supervulcanoes.

↓

Radiation cycles seem one likely cause – but no obvious explanation of these cycles have been generally accepted.
Article in *Nature* (1976) "Variation of the earth's orbit: pacemaker of the ice age".

History of Western Knowledge of Rain and Clouds (Nephology): 1. Outline

Not Chronologically Proportional

Many ancient philosophers wrote works on natural history and philosophy, which often could contain discussions of atmospheric phenomena. According to Diogenes Laertius, the only surviving ancient history of philosophy, two philosophers, Theophrastus and Posidonius, apart from Aristotle who wrote *Meteorologica*, wrote works entitled *Meteorology*. He does not actually mention Aristotle's *Meteorologica*, but lists another work by him called *Prognostics of Storms*, in one book.

More Naturalistic — Less Naturalistic

PreSocratics, c. 600-370 BC
Thales
Clouds are form of water rising and sinking in the air (atmosphere)
Anaximander and **Anaxagoras**
Wind consists of flowing air, and lightning is created by friction within clouds
Democritus. An atomic and naturalistic theory of clouds. Clouds and rain are water atoms which rises from the earth, are carried south by winds and then rain down (presenting a water cycle).

Aristotle, *Meteorologica*
Uses the four elements and damp-dry and warm-cold to explain how everything searches its telos (place, purpose).
Clouds and rain are due to warm and damp air rises from the earth. Warm and damp air also explain the cause why clouds move around the earth. Clouds are created in medium height in the atmosphere (too high there is too much 'fire' and too low too much reflected light).

Epicurus
Continues a highly naturalistic explanation of clouds and weather.
↓
Lucretius

Middle Ages
More religious and moral explanations of clouds and weather.
Isidorus of Sevilla
Bede, *De Natura Rerum*
et al

Shortly after 1600 many new instruments were discovered, including the thermometer and barometer

Descartes, c. 1637
Gives a highly naturalistic and mechanistic explanation of clouds

Luke Howard, 1802
Gives names and divides clouds into 3 and 7 main types: cumulus, stratus and cirrus.
Modern understanding of clouds.

Taxonomy of Clouds: 1. Outline

Historically, when Luke Howard in 1802/03 first systematized and gave name to clouds, there were only three main types (or genera) of clouds, divided on the basis of shape (or form), but also divided into three different heights in the atmosphere (troposphere).

Clouds are formed when relatively moist air rises (and cools). This is generally caused by either heated air (by the sun) rising (in the summer), geographical formations like mountains and hills forcing the air upwards or by fronts (warmer or colder air pushing from the side on the present air, when the cooler air always take the lower position).

Name of Cloud Type / Height in the Atmosphere	Cirrus (Hairy or feathery) Latin for hair	Stratus (layered) Latin for layer, extended	Cumulus (heap, pile) Latin for pile
High Over 5 000 to 13 000 m.	**Cirrus**	**Cirrostratus**	**Cirrocumulus**
Medium 2 -7 000 m.	-	**Altostratus**	**Altocumulus**
Low Bottom of clouds less than 2 000 m.	-	**Stratus**	**Cumulus**

Taxonomy of Clouds: 2. Overview

Today ten main forms (genera) are recognized in the standard *International Cloud Atlas*. The original seven types (seven rather than nine due to that there are no medium or low cirrus clouds) are combined with two which give rain or precipitation, *nimbus* (but also altostratus can cause steady precipitation and stratus sometimes drizzle and cumulus can give rise to showers) and a combination of stratus and cumulus, stratocumulus (which is actually the most common form of cloud, but it is mainly formed over the sea so not seen as frequently on land). These ten can be further specified into a number of different species, but those are generally less important.

Name of Cloud Type / Height in the Atmosphere	Cirrus (Hairy or feathery) Latin for hair	Stratus (layered) Latin for layer, extended	-	Cumulus (heap, pile) Latin for pile	Nimbus (Rain, Percipitation) Latin for rain	
High Over 5 000 to 13 000 m. Generally ice crystals	**Cirrus**	**Cirrostratus**	**(Stratocumulus)**	**Cirrocumulus**	-	**(Cumulonimbus)**
Medium 2 -7 000 m. Alto is Latin for high. Mixture of liquid water and ice crystals	-	**Altostratus**	**(Stratocumulus)**	**Altocumulus**	**(Nibostratus)**	**(Cumulonimbus)**
Low Bottom of clouds less than 2 000 m. Mixture of liquid water and ice crystals	-	**Stratus**	**Stratocumulus**	**Cumulus**	**Nimbostratus** (Steady rain)	**Cumulonimbus** (Thunderstorms)

Cloud types with name in parenthesis are normally not visible from the ground.

The Rock Cycle (Due to Plate Tectonics)

There are very many different sorts of rocks and minerals, but fundamentally they can be divided into three main types; igneous, metamorphic and sedimentary.

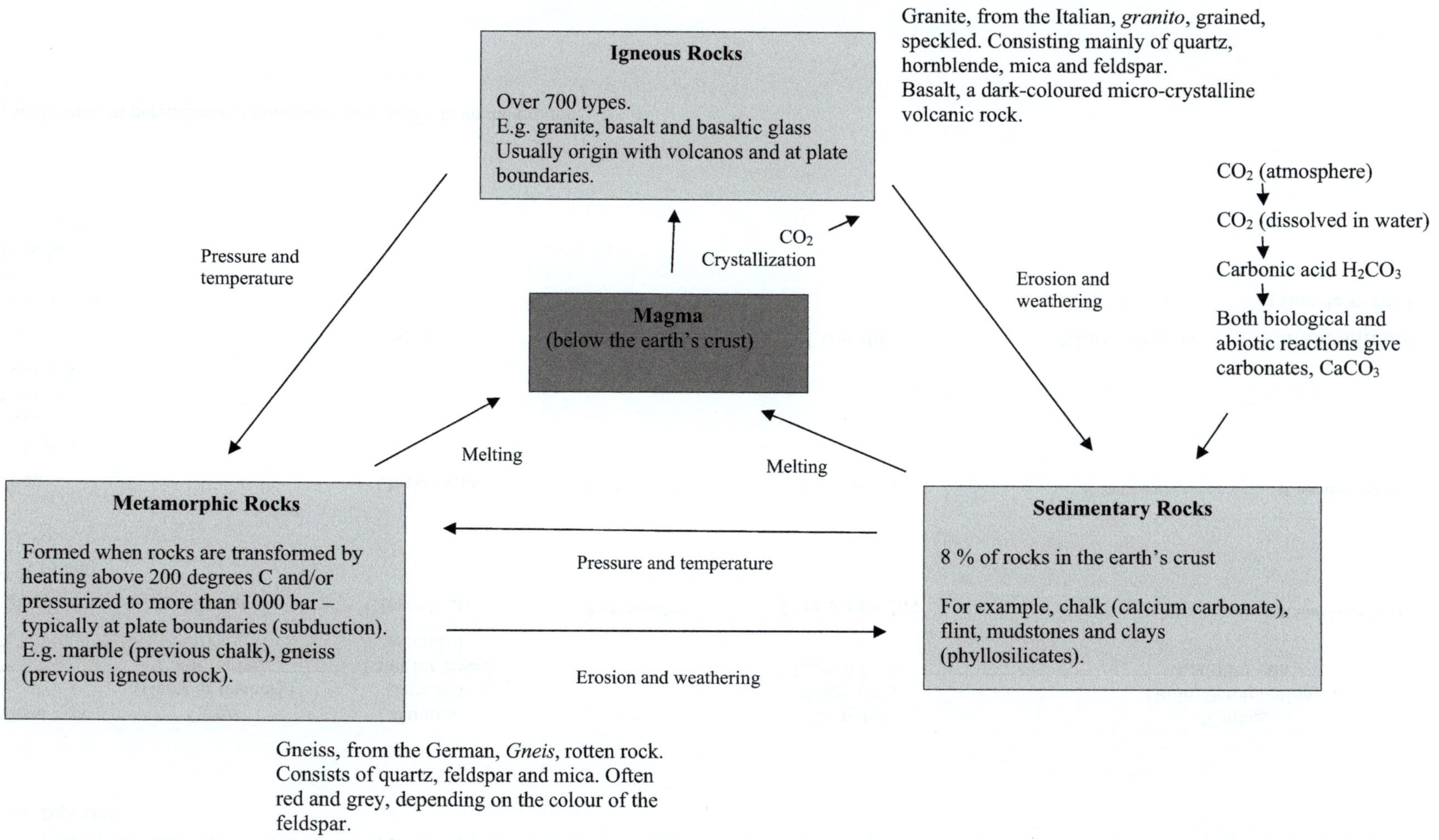

Index

T. H. Brobjer, *Atlas of the History of Modern Science 1500-2020*, https://doi.org/10.1007/978-3-031-76627-5

9783031766268